01-WSM-1

NAVIGATING THE TMDL PROCESS: METHOD DEVELOPMENT FOR ADDRESSING NARRATIVE CRITERIA

by:

Principal Investigators
Paul L. Freedman, P.E., DEE
David W. Dilks, Ph.D.
Limno-Tech, Inc.

Hans P. Holmberg, P.E.
Penelope E. Moskus
Limno-Tech, Inc.

Graham McBride
Chris Hickey
David G. Smith, Ph.D.
NIWA

Peter L. Striplin
Striplin Environmental Associates

2003

Water Environment Research Foundation

The Water Environment Research Foundation, a not-for profit organization, funds and manages water quality research for its subscribers through a diverse public-private partnership between municipal utilities, corporations, academia, industry, and the federal government. WERF subscribers include municipal and regional water and wastewater utilities, industrial corporations, environmental engineering firms, and others that share a commitment to cost-effective water quality solutions. WERF is dedicated to advancing science and technology addressing water quality issues as they impact water resources, the atmosphere, the lands, and quality of life.

For more information, contact:
Water Environment Research Foundation
635 Slaters Lane, Suite 300
Alexandria, VA 22314-1177
Tel: (703) 684-2470
Fax: (703) 299-0742
www.werf.org
werf@werf.org

This report was co-published by the following organizations. For non-subscriber sales information, contact:

Water Environment Federation
601 Wythe Street
Alexandria, VA 22314-1994
Tel: (800) 666-0206
Tel: (703) 684-2452
Fax: (703) 684-2492
www.wef.org
pubs@wef.org

IWA Publishing
Alliance House, 12 Caxton Street
London SW1H 0QS, United Kingdom
Tel: +44 (0) 20 7654 5500
Fax: +44 (0) 20 7654 5555
www.iwapublishing.com
publications@iwap.co.uk

Library of Congress Catalog Card Number: 2003106699

WERF ISBN: 1-893664-89-9 IWAP ISBN: 1-84339-671-8 WEF ISBN: 1-57278-224-2

Limno-Tech, Inc.

The research on which this report is based was funded, in part, by the United States Environmental Protection Agency through Cooperative Agreement No. CR-827345-4 with the Water Environment Research Foundation (WERF). Unless an EPA logo appears on the cover, this report is a publication of WERF, not EPA. Funds awarded under the Cooperative Agreement cited above were not used for editorial services, reproduction, printing, or distribution.

ACKNOWLEDGMENTS

Report Preparation

Principal Investigators:
Paul L. Freedman, P.E., DEE
David W. Dilks, Ph.D.
Limno-Tech, Inc.

Project Team:
Hans P. Holmberg, P.E.
Penelope E. Moskus
Limno-Tech, Inc.

Graham B. McBride
Christopher W. Hickey, Ph.D.
David G. Smith, Ph.D.
National Institute of Water and Atmospheric Research (NIWA)

Peter L. Striplin
Striplin Environmental Associates

Project Subcommittee

Chair and WERF Research Council Liaison:
Norman E. LeBlanc, *Hampton Roads Sanitation District*

WERF Research Council Liaison:
Philip B. Dorn, Ph.D., *Shell Global Solutions*

Larry W. Ausley, *North Carolina Department of Environment and Natural Resources, Division of Water Quality*

Donald J. Brady, Ph.D., *U.S. Environmental Protection Agency, Office of Water*

Robert A. Hoke, Ph.D., *DuPont Company*

Phillip Jennings, *U.S. Environmental Protection Agency, Region 6*

Kenneth H. Reckhow, Ph.D., *Duke University; Water Resources Research Institute, North Carolina State University*

Water Environment Research Foundation Staff

Director of Research: Daniel Woltering, Ph.D.
Project Managers: Dean E. Carpenter, D.Env.
Eugenia Bruno Hart

ABSTRACT AND BENEFITS

Abstract:

Narrative water quality criteria are an integral component of state water quality standards but they pose special challenges when incorporated into the TMDL process. The TMDL process is typically a quantitative activity, defining numeric allowable loads necessary to meet a numeric water quality objective. Narrative criteria confound development of quantitative TMDLs, due to the absence of these numeric criteria. The objective of this research was to improve the methods for addressing narrative water quality criteria in the TMDL process. Based on a comprehensive review of more than 120 narrative criteria TMDLs, the authors summarize the range of challenges associated with the TMDL process for the various types of narrative criteria, and provide guiding principles to help address these challenges. The guiding principles should significantly improve the evaluation of water bodies for impairment of a designated use based on non-attainment of narrative water quality criteria, resulting in more accurate 303(d) lists with a greater focus on those waterbodies truly impaired. Adherence to these guiding principles also should improve the TMDL development process, resulting in a greater degree of success in attaining designated uses when a TMDL is implemented.

Benefits:

- The research provides a comprehensive review of 120 TMDLs developed to address impairments of the designated use of water bodies based on non-attainment of narrative water quality criteria. A broad range of TMDLs were reviewed, including TMDLs developed to address toxicological, ecological, and aesthetic impairments. Strengths and weaknesses in the TMDLs are documented as they pertain to defining impairment, determining causes of impairment, and setting TMDL target measures and levels.
- The report proposes a series of guiding principles to improve the process for developing TMDLs for narrative criteria. These guiding principles are targeted for the TMDL developer, but they also should assist stakeholders involved in the TMDL review process.
- The guiding principles are designed to increase the likelihood of attaining designated uses when a TMDL is implemented by increasing the rigor and accuracy of determination of causes of impairment and the selection of appropriate TMDL target levels.
- This report presents case examples to illustrate how the guiding principles were or were not followed in the development of actual TMDLs.

Keywords: Impairment, water quality standards, 303(d) list

TABLE OF CONTENTS

LIST OF TABLES

LIST OF FIGURES

LIST OF ACRONYMS

BAF	Bioaccumulation Factor
BBN	Bayesian Belief Networks
BPJ	Best Professional Judgment
CALM	Consolidated Assessment and Listing Methodology
EPT	Ephemeroptera, Plecoptera and Trichoptera (mayflies, stoneflies and caddisflies)
IBI	Index of Biotic Integrity
NALMS	North American Lake Management Society
NPDES	National Pollutant Discharge Elimination System
NRC	National Research Council
NA	Not Applicable
NE	No Evidence
PAH	Polycyclic Aromatic Hydrocarbons
SETAC	Society of Environmental Toxicology and Chemistry
TIE	Toxicity Identification and Evaluation
TMDL	Total Maximum Daily Load
TSS	Total Suspended Solids
TSI	Trophic Status Index
U.S. EPA	United States Environmental Protection Agency
WERF	Water Environment Research Foundation
WET	Whole Effluent Toxicity
WQCD	Water Quality Control Division
WQS	Water Quality Standard

Executive Summary

ES.1 The Need for this Research

Section 303 of the Clean Water Act requires states or the U.S. Environmental Protection Agency to establish numeric pollutant loading targets (termed total maximum daily loads, or TMDLs) for waters that are not meeting water quality standards on the basis of technology-based controls on point sources. Many water bodies are listed as impaired because of non-attainment of narrative, rather than numeric, water quality criteria. In such cases, the development of a TMDL presents special challenges; the determination of impairment and development of TMDLs is typically a quantitative activity, defining numeric allowable loads necessary to meet a numeric water quality objective. Narrative criteria may confound the determination of impairment and development of quantitative TMDLs because of the absence of specific numerically defined objectives.

The objectives of this research are to survey the "state of the practice" for TMDL development for narrative criteria, summarize the range of challenges encountered in these TMDLs, and recommend a set of guiding principles to improve key aspects of the TMDL process for narrative criteria.

The guiding principles are targeted for the TMDL developer, but they also should be helpful to stakeholders as part of the TMDL review process. Because of the many challenges posed by narrative criteria TMDLs, affected parties should consider active involvement in the TMDL development process to ensure that these principles are followed. This involvement can range from external review to working cooperatively with the designated regulatory authority to conduct the TMDL.

ES.2 Summary of Review

The first step toward reaching the research objective was to characterize the current "state of the practice" of TMDL development for narrative criteria. Hundreds of TMDLs already have been conducted for waterbodies listed as impaired because of non-attainment of narrative criteria. These studies provide a history of the challenges encountered in addressing narrative criteria and potential means to address those challenges.

The narrative TMDLs reviewed as part of this research spanned a wide spectrum of narrative criteria types, and included more than 120 narrative TMDLs. Appendix A of the report contains a complete listing of the TMDLs reviewed. Impairment types for narrative criteria were grouped into three categories, according to the language in the narrative criterion responsible for the listing:

- Toxicological (e.g., "No toxics in toxics amounts");
- Ecological (e.g., "Maintain the biological integrity of the waters of the state"); and
- Aesthetic (e.g., "No nuisance amounts of …").

Within these categories, TMDLs were reviewed in the context of three steps in the TMDL process that pose unique challenges when addressing narrative criteria:

- Determination of Impairment: Defining impairment because of non-attainment of narrative criteria;
- Problem Identification: Determining causes of impairment; and
- Target Analysis: Setting TMDL target measures and levels.

The review identified frequent and/or important problem areas for each of the three categories of narrative criteria and in each of the three steps of the TMDL process posing challenges to addressing narrative criteria (summarized in Table ES-1).

Table ES-1. Summary of Challenges Identified in the Review of Narrative Criteria TMDLs.

Categories of Narrative Criteria	Steps in the TMDL Process Presenting Challenges Unique to Narrative Criteria		
	Defining Impairment	**Determining Causes of Impairment**	**Setting TMDL Target Measures and Levels**
Toxicological	Use of evaluated data to presume impairment Interpreting ambient toxicity data; and Listing based on fish consumption advisory	Determining cause based on ambient toxicity data	Setting TMDL targets in water column for complex chemical speciation and uptake
Ecological	Use of surrogate measures may not be reflective of criteria Selection of impairment endpoint inappropriate for the water body	Complex and poorly understood cause-and-effect mechanisms	Selecting target measures reflective of criteria and also able to be linked to controls
Aesthetic	Converting perceptions to levels	—	Converting perceptions to levels

The results of the review found that toxicological TMDLs were conducted primarily as a result of whole effluent toxicity tests, and human health concerns from fish tissue contamination. Even with a focused effort by the project team to identify TMDLs based on ambient toxicity data, only one of the TMDLs compiled was based on actual ambient toxicity data. When defining impairment, toxicity TMDLs were identified as posing obstacles when using evaluated data to presume impairment and also when interpreting ambient toxicity data. Determining the causes or sources of impairment based on ambient toxicity data also was a major challenge.

Another area of concern was listing of a water body as impaired *strictly* because of the presence of fish consumption advisories; depending on the cause of the advisory, it may, or may not, indicate impairment. Finally, the results of the review indicated that it is extremely difficult to set TMDL target levels in the water column that will result in acceptable fish tissue concentrations for some chemicals, such as mercury, that undergo complex speciation in the natural environment and uptake through the food chain.

Narrative criteria TMDLs in the Ecological category consisted primarily of sediment/habitat impairments, with some eutrophication TMDLs also falling into this category. The review revealed that impairment often was defined using surrogate measures of the designated use for this category of TMDLs. Surrogate measures present a challenge, because they may poorly reflect the objective of the narrative criterion. The review also noted that impairment decisions sometimes were based on an inappropriate level for the water body. For example, the target level of a macrobenthos index for a channelized urban stream should reflect the local habitat quality, and would be different than the target level for a pristine watershed. Determining the causes of impairment presented significant challenges for narrative criteria TMDLs in this category because of complex and typically poorly understood cause-and-effect relationships between stressors and ecological responses. Finally, selecting TMDL target measures that were both reflective of the narrative criterion and able to be linked to TMDL controls proved difficult for this category.

Finally, TMDLs in the Aesthetic category were developed for water bodies impaired by turbidity, debris, or the consequences of eutrophication. The primary challenge in these TMDLs was determining appropriate levels for defining impairment and setting TMDL targets. The difficulty here is determining what quantity of a given constituent/characteristic constitutes a nuisance level.

ES.3 Summary of Guiding Principles

A wide range of existing narrative criteria TMDLs was evaluated to define the most common issues, and basic guiding principles were developed to address the most common or challenging issues unique to narrative criteria. Narrative criteria cover a diverse spectrum of water quality objectives, and approaches for the development of TMDLs to address this diversity are numerous. This study does not provide detailed guidance on all aspects of addressing narrative criteria in the TMDL process. However, it does provide general guiding principles for TMDL developers and stakeholders to follow. The guiding principles developed as part of this research are summarized in Table ES-2 and in a set of checklists in Appendix B. Case examples are provided to demonstrate existing TMDLs that do (or do not) adhere to these principles.

Adherence to the guiding principles should significantly improve the evaluation of water bodies for impairment of a designated use based on non-attainment of narrative water quality criteria, resulting in more accurate CWA 303(d) lists and a greater focus on those water bodies that truly are impaired. Adherence to these guiding principles also will contribute to improvements in the TMDL development process, resulting in a greater degree of success in attaining designated uses when a TMDL is implemented. This would be accomplished through more accurate determination of the causes of impairment, and by setting TMDL target levels that, if achieved, will result in attainment of the designated use.

Table ES-2. Summary of Guiding Principles.

Guiding Principles for Defining Impairment	
For a Consistent Decision Process	
1	States should develop protocols
2	Consider all relevant available protocols
3	Document all rationale
4	Use an intermediate listing category if impairment is inconclusive
5	Investigate the scientific basis of fish consumption advisories
For Selection of Measures and Levels of Impairment	
1	Link the measures to the designated use
2	Use well-accepted techniques to determine the measure
3	Determine the expected variability of the measurement
4	Measures should be consistent with existing data
5	Measure should be readily understood.
6	Consider multiple measures – weight of evidence approach
7	The threshold level must be relevant to the specific designated use and local condition
For Data Interpretation	
1	Address variability in measurement
2	Consider geographic and temporal relevance
3	Consider all potential causes
For Determining Aesthetic Impairment	
1	Assess potential causes
2	Develop measurement procedures
3	Take into consideration local input

Guiding Principles for Determining Causes of Impairment	
For Listing Candidate Causes	
1	Develop a wide initial list of candidate sources and causes
2	Develop a conceptual model indicating links between causes and impairment
For Analyzing the Evidence	
1	Analyze available data and literature to refine the conceptual model
2	Assign uncertainties to the model and use the model to guide data collection
For Identifying Potential Causes	
1	Eliminate causes that fail logical tests
2	Subject all remaining causes to diagnostic analysis
3	Perform a strength-of-evidence analysis
4	Apply the Bradford-Hill criteria to the strength-of-evidence analysis
5	Compare the relative strengths of evidence using a weighted matrix for causes

Table ES-2. Summary of Guiding Principles, continued.

Guiding Principles for Setting TMDL Target Measures and Levels	
For Selecting Target Measures	
1	When cause-effect is well understood select direct measures of environmental response
2	When cause-effect is poorly understood select surrogate measures to serve as TMDL targets
3	A direct TMDL target measure should be the same as used to determine impairment
4	Use the most protective measure as the TMDL target when multiple direct measures are used
5	Surrogate targets must be linked to source controls showing significant causative factors
For Setting Target Levels	
1	Historic monitoring data provide the most direct means of defining TMDL target levels
2	A reference watershed must reflect attainable conditions in the target watershed
3	A reference watershed must have targets with similar causative mechanisms

CHAPTER 1.0

INTRODUCTION

The objectives of the 1972 Water Pollution Control Act, more commonly referred to as the Clean Water Act (CWA), include achieving water quality that provides for protection and propagation of a balanced population of fish, shellfish, and wildlife, and supports recreation in and on the water. To help implement these objectives, CWA Section 303 requires that each state develop water quality standards that include a designated use, numeric or narrative criteria to support that use, and an anti-degradation policy.

The listing of a waterbody as impaired is handled within the CWA Section 303(d) process and is usually initiated by the state agency. Impairment is determined by nonattainment of state water quality standards that are subject to approval by the U.S. Environmental Protection Agency (U.S. EPA). The state's list of impaired waters also is reviewed by U.S. EPA for approval. Once a waterbody is listed as impaired, a process begins for identifying the allowable loadings for specific pollutants that the waterbody can receive and still attain water quality standards. This is known as a total maximum daily load, or TMDL.

Many waterbodies have been listed as impaired and have had TMDLs developed because of non-attainment of narrative criteria. While narrative criteria are an integral component of each state's water quality standards, such criteria can pose special challenges when incorporated into the TMDL process. The determination of impairment and development of TMDLs typically is a quantitative activity, defining numeric allowable loads necessary to meet a numeric water quality objective. Narrative criteria can confound the determination of impairment and development of quantitative TMDLs because of the absence of specific numerically defined objectives.

The objective of this research is to improve the methods for addressing narrative water quality criteria in the TMDL process. This chapter introduces the concept of narrative water quality criteria (Section 1.1) and TMDLs (Section 1.2), highlights challenges associated with narrative criteria in the TMDL process (Section 1.3), and describes the general intent and approach of this research (Section 1.4).

1.1 Narrative Water Quality Criteria

To support the development of water quality standards by states, the U.S. EPA developed the Water Quality Standards Handbook (U.S. EPA, 1994). For most waterbodies and many pollution conditions, the U.S. EPA or states have developed numeric criteria for specific pollutants. When numeric criteria are developed, however, it is not possible to anticipate all pollutants or endpoints of concern, nor is it possible to derive some types of numeric criteria that will apply generally across all waterbodies. Often data are insufficient to develop site-specific numeric water quality criteria at the time of water quality standards adoption. Recognizing these situations, standards-setting authorities adopt narrative criteria to ensure full support of the designated uses. Narrative criteria can descriptively accomplish what numeric criteria, in many cases, cannot account for quantitatively at the time water quality standards are adopted (U.S. EPA, 2000a). Therefore, certain pollution conditions may be governed only by narrative criteria with or without promulgated quantitative measures. Narrative criteria are defined in U.S. EPA's Water Quality Standards Handbook (1994) as follows:

> States are required to adopt water quality criteria, based on sound scientific rationale, that contain sufficient parameters or constituents to protect the designated use. EPA believes that an effective state water quality standards program should include both parameter-specific (e.g., ambient numeric criteria) and narrative approaches....Narrative criteria are statements that describe the desired water quality goal, such as the following:
>
> All waters, including those within mixing zones, shall be free from substances attributable to wastewater discharges or other pollutant sources that:
> (1) Settle to form objectionable deposits;
> (2) Float as debris, scum, oil, or other matter forming nuisances;
> (3) Produce objectionable color, odor, taste, or turbidity;
> (4) Cause injury to, or are toxic to, or produce adverse physiological responses in humans, animals, or plants; or
> (5) Produce undesirable or nuisance aquatic life (54 F.R. 28627, July 6, 1989).
>
> EPA considers that the narrative criteria apply to all designated uses at all flows and are necessary to meet the statutory requirements of section 303(c)(2)(A) of the CWA.

Narrative criteria are qualitative descriptions of the conditions of the waterbody necessary to support the designated uses, while numeric criteria quantitatively define the permissible level of a specific pollutant to maintain protection of a designated use. Thus, narrative water quality criteria establish a basic foundation for supporting designated uses while numeric water quality criteria provide a specific quantitative translation of the necessary level of protection.

1.2 303(d) Listing and Total Maximum Daily Loads (TMDLs)

Section 303(d) of the Clean Water Act requires states to identify waters that do not attain water quality standards after the implementation of technology-based effluent limits at all point sources. States are required to develop lists of these impaired waters, commonly referred to as the 303(d) list. Section 303(d) then requires that a total maximum daily load (TMDL) be established at a level necessary to attain the applicable water quality standard. A TMDL is defined as the greatest amount of pollutant loading a waterbody can receive and still maintain water quality standards while considering a margin of safety. Recent emphasis has been put on the TMDL program to serve as a primary mechanism to provide the attainment of water quality standards for impaired waterbodies.

Many of the terms used to describe the determination of impairment and development of TMDLs (e.g., endpoint, target, indicator) have been used interchangeably and inconsistently, and are a potential source of confusion. For example, the U.S. EPA (1999a) sediment TMDL guidance defines endpoint as follows:

> **Endpoint.** An endpoint (or indicator/target) is a characteristic of an ecosystem that may be affected by exposure to a stressor. Assessment endpoints and measurement endpoints are two distinct types of endpoints commonly used by resource managers. An assessment endpoint is the formal expression of a valued environmental characteristic and should have societal relevance (an indicator). A measurement endpoint is the expression of an observed or measured response to a stress or disturbance. It is a measurable environmental characteristic that is related to the valued environmental characteristic chosen as the assessment endpoint. The numeric criteria that are part of traditional water quality standards are good examples of measurement endpoints (targets).

This definition highlights the range of interchangeably used terms (i.e., endpoint and indicator/target) and seemingly similar terminologies such as *assessment endpoints* and *measurement endpoints*. For the purpose of this research report, the following terms are defined:

- *endpoint*: a generic term, encompassing a range of essential ecological characteristics (e.g., habitat, biological integrity) that may include quantifiable measures.
- *target:* similar to endpoint, but implying greater specificity and relating to a quantifiable characteristic. A target is comprised of a measure and a level – see below.
- *measure:* the specific characteristic or feature being evaluated. Examples for target measures for habitat would be percentage of eroded streambank; a measure for biological integrity would be the use of the macroinvertebrate index of biotic integrity (IBI).
- *level:* a numeric value associated with a specific measure. An example of a level would be no more than 10% eroded streambank or a specific IBI score.

The initiation of the TMDL process (Figure 1-1) involves making a determination of impairment. This decision is based on the water quality standards that apply to the waterbody in question and the available data characterizing that waterbody. If this comparison of data to standards results in a determination that the waterbody is meeting standards, the waterbody is not included on the 303(d) list. The process does not end at this point, however, because data collection should continue and the determination of impairment should be reevaluated on an appropriate time interval. On the other hand, if the comparison of data to standards results in a determination that the waterbody is impaired, the waterbody is placed on the 303(d) list. States then rank waterbodies on the 303(d) list by priority for TMDL development. Once the TMDL development process is initiated for a waterbody, U.S. EPA (1999a, 1999b) guidance lists six core steps to follow:

- **Problem Identification:** Identify key factors and background information that describe the nature of the impairment. An essential component of Problem Identification is making a determination of the causes of impairment.

- **Target Analysis—Identification of Water Quality Indicators and Targets:** Identify TMDL target measures and set levels designed to reflect the attainment of water quality standards.
- **Source Assessment:** Characterize the current sources of the causes of impairment in terms of type, magnitude, and location.
- **Linkage Analysis - Link Sources to Targets and Estimate Loading Capacity:** Develop a quantitative relationship between the selected TMDL target measures and the identified sources, and identify the maximum allowable loads that will result in meeting the TMDL target level, including a margin of safety.
- **Allocation:** Specify the allowable magnitude of the contributing sources, by source category.
- **Development of an Implementation, Monitoring and Review Plan**

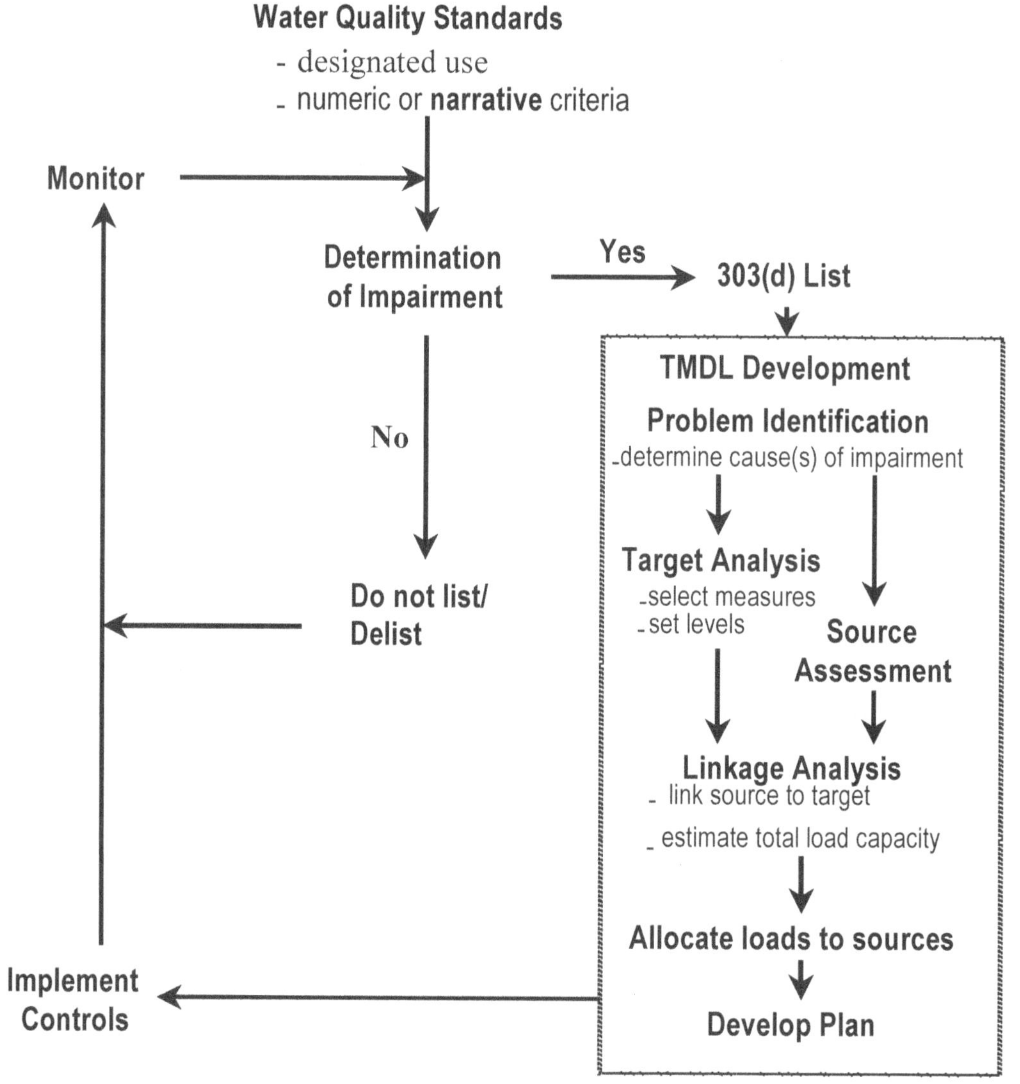

Figure 1-1. Flowchart of Steps in the TMDL Process.

1.3 Challenges of Addressing Narrative Criteria in the TMDL Process

Two basic qualities of narrative criteria present challenges in the TMDL process: 1) absence of specific quantitatively defined water quality objectives, or numeric criteria, and 2) unclear or multiple causes of impairment. These two qualities pose challenges unique to narrative criteria at three specific steps in the TMDL process:

- Determination of Impairment
- Problem Identification
- Target Analysis

Once numeric targets are established for a narrative TMDL in the Target Analysis step, the remaining steps of the TMDL process are similar to TMDLs developed for numeric criteria. As discussed below, the challenges of addressing narrative criteria in each of these three steps of the TMDL process are inter-related.

1.3.1 Determination of Impairment

The TMDL process begins with a determination of impairment for a waterbody based on its non-attainment of water quality standards. For narrative criteria, the absence of a specific quantitative threshold poses an immediate challenge. Specification of numeric translators that reflect some narrative criteria in quantitative terms can be a difficult process, involving numerous scientific and policy decisions. In the absence of numeric translators, best professional judgment must be used to define impairment. If expertise is lacking, judgments can be poor, but even experts can differ in judgment without specific numeric criteria to guide decisions. The subjective nature of this type of site-specific judgment leads to inconsistent and sometimes inaccurate determinations of impairment.

Data representing a direct measure of the narrative criteria are not always available or easily obtained; therefore, surrogate measures or evaluated data are often used in assessing attainment of narrative criteria. (In this report, evaluated data refers to an indirect evaluation of water quality through the use of predictive modeling.) Surrogate measures and evaluated data are not always clearly representative of the objectives of a narrative criterion and may not accurately reflect whether impairment exists. Also, the types of data used for assessment of narrative criteria may be complex and often related to biological endpoints. These types of data require more rigorous interpretation than the chemical-specific measurements used when assessing numeric criteria.

1.3.2 Problem Identification

This step in the TMDL process identifies key factors and background information that describe the nature of the impairment, including the causes of impairment. Determining the causes of impairment based on non-attainment of narrative criteria presents challenges. As mentioned above, narrative criteria often are related to biological endpoints (e.g., fish populations), which can vary widely over time and space and can be impacted by a diverse range of stressors. The stressors causing impairment may be human-induced or natural, including floods or drought. The TMDL analyst is required to make a determination of the causes of the depressed biological condition, noting that TMDLs are required only for human-induced impairment. If the determination of the causes of impairment is done incorrectly, resources may be spent addressing the wrong issues. If one of multiple causes of impairment is not included in the TMDL, attainment of the narrative criterion may not be achieved. This type of judgmental determination typically is not required for chemical specific numeric criteria.

1.3.3 Target Analysis

Finally, setting specific target measures and levels in the Target Analysis step of TMDL development is challenging when the impairment is based on nonattainment of narrative criteria. When addressing numeric criteria in the TMDL process, the measure used for determination of impairment and the TMDL target are typically both defined as the numeric criterion.

The difficulty in this step involves unclear (and potentially multiple) stressor-response relationships that are common to biological endpoints reflected by narrative criteria, as discussed above. TMDLs are more easily developed when there is a known and quantifiable link between causes/sources to be controlled and responses desired. Figure 1-2 (from NRC, 2001) shows the locations along the causal chain between stressors and responses where TMDL targets can be specified. For many biological responses, these relationships are poorly understood, resulting in high uncertainty in many of the linkages (reflected by the arrows in Figure 1-2). Without this knowledge, it is difficult to calculate a TMDL linking causes (sources or stressors) to effects (water quality response).

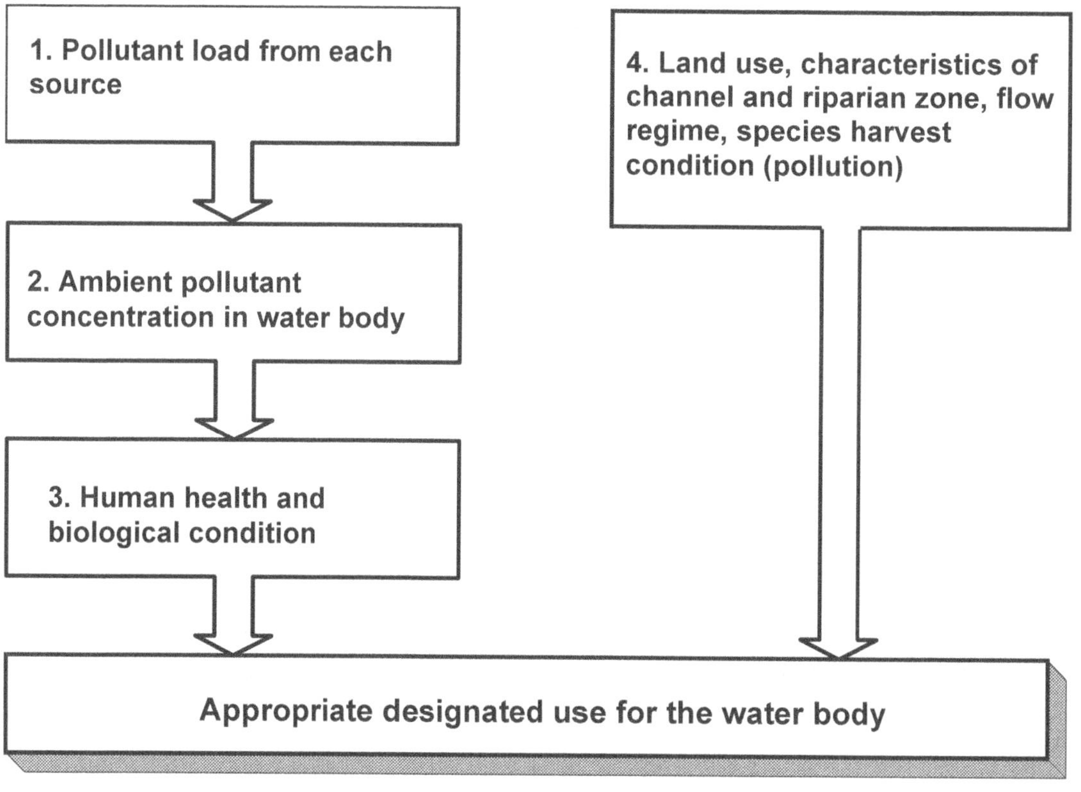

Figure 1-2. Types of water quality criteria and their position relative to designated uses.
(Reprinted with permission from NRC, 2001. Copyright 2001 National Academy of Sciences.)

The TMDL target can be located at any point along the causal chain shown in Figure 1-2. For an effective TMDL, the target must be directly linked to the causes (e.g., effluent limits, sources of impairment other than chemical pollutants), to determine the needed controls on the sources. The target also must be directly linked to the designated use or narrative criteria to support that use. The uncertain linkages associated with narrative criteria, therefore, pose a dilemma in setting TMDL targets. The target can be located near the source end of the chain, where it can be directly linked to TMDL controls but potentially poorly reflect the objectives of the narrative criteria. The alternative is to locate the TMDL target closer to the response end of the chain, which more accurately reflects the objectives of the narrative criteria but provides uncertain linkage to the required TMDL controls. Thus, for narrative TMDLs, it can be difficult to select a target measure, let alone set a target level.

1.4 Research Intent

The challenges associated with narrative criteria in the TMDL process are known in general terms and have been reported on an anecdotal basis (e.g., Pott et al., 2002), but, prior to this research, no thorough review of the use of narrative criteria in TMDLs or guidance on how to address narrative criteria in the TMDL process has been performed. U.S. EPA has developed TMDL guidance for two impairments commonly associated with narrative criteria, excess sedimentation (U.S. EPA, 1999a) and nutrients (U.S. EPA, 1999b). U.S. EPA is also developing guidance on the 303(d) listing process that will address narrative criteria to some extent (U.S. EPA, 2001a). While some guidance exists or is being developed, it addresses only a subset of the challenges associated with addressing narrative criteria in the TMDL process. The intent of this research is to define the range of issues associated with all types of narrative criteria and provide guiding principles to help address these challenging issues.

A wide range of narrative criteria exists, as well as a variety of approaches for developing TMDLs to address impairments of narrative criteria. It is beyond the scope of this research to provide detailed guidance on all aspects of addressing narrative criteria in the TMDL process. The approach taken, therefore, was to review a wide range of existing narrative criteria TMDLs to better define the most common challenges, and then to provide basic guiding principles to address these challenges. Adherence to these principles should significantly improve the evaluation of waterbodies for impairment of a designated use based on nonattainment of narrative water quality criteria, resulting in more accurate 303(d) lists with a greater focus on those impaired waterbodies. Also, adherence to these guiding principles will contribute to improvements in the TMDL development process, resulting in a greater degree of success in attaining designated uses when a TMDL is implemented.

CHAPTER 2.0

REVIEW OF NARRATIVE CRITERIA TMDLS

2.1 Introduction

The objective of this research is to improve the methods for addressing narrative criteria in the TMDL process. The first step towards reaching that objective was to characterize the current "state of the practice" of TMDL development for narrative criteria. Hundreds of TMDLs already have been conducted for waterbodies listed as impaired because of nonattainment of narrative criteria. These TMDLs provide a rich history of the challenges encountered in addressing narrative criteria and potential means to address those challenges. This section describes the review of TMDLs for waterbodies listed as impaired because of nonattainment of narrative criteria, and identifies the major issues unique to narrative criteria TMDLs in need of additional guidance.

The review of TMDLs considered a wide spectrum of narrative criteria types, and included more than 120 TMDLs developed for narrative criteria. The distribution of types of TMDLs chosen for review was selected to roughly match the distribution of impairment types found for all 303(d)-listed waterbodies. Most of the TMDLs were for excess sedimentation and nutrient impairments, with a representative number of additional TMDLs reviewed for other impairments. Impairment types for narrative criteria were subsequently grouped into three categories for purposes of this review, according to the language in the narrative criterion responsible for the listing:

- Toxicological (e.g., "No toxics in toxic amounts");
- Ecological (e.g., "Maintain the biological integrity of the waters of the State"); and
- Aesthetic (e.g., "No nuisance amounts of …").

Within these categories, TMDLs were reviewed in the context of the three steps in the TMDL process identified in Section 1.3 as posing unique challenges when addressing narrative criteria:

- Determination of Impairment: Defining impairment because of nonattainment of narrative criteria;
- Problem Identification: Determining causes of impairment; and
- Target Analysis: Setting TMDL target measures and levels.

2.2 The Review Process

TMDLs were reviewed in two phases. A preliminary review was conducted using 28 readily available narrative criteria TMDLs contained in the database compiled by the WERF project "Critical Assessment of the TMDL Process: Evaluation and Improvement" (Limno-Tech et al., 2003). The findings of the preliminary review provided valuable information on the range of impairments for which narrative TMDLs had been developed and the nature of the challenges associated with narrative TMDLs. These findings were then used to guide the compilation of an expanded database of over 100 narrative TMDLs and to design the structure for the detailed review of these TMDLs. The components of the review process are grouped as follows:

- Preliminary Review;
- Compilation of Additional Narrative Criteria TMDLs; and
- Structure of Narrative TMDL Review.

2.2.1 Preliminary Review

The preliminary review focused on approved TMDLs compiled under another WERF project, "Critical Assessment of the TMDL Process: Evaluation and Improvement" (Limno-Tech et al., 2003). From this database of 176 TMDLs, 28 were identified as being developed for narrative criteria. The information collected for each of those TMDLs included the TMDL development document, any supporting documents on data, tools and administrative processes, and the name of the TMDL developer. During the preliminary review, TMDLs were characterized with respect to the following:

- Impairment types;
- The degree to which impairment was documented; and
- The process by which numeric targets were developed for narrative criteria.

The first review topic was intended to document the range of impairment types for which narrative criteria TMDLs were developed, and the relative frequency of each impairment type. This topic of review was intended to provide general background information, as well as to determine the availability of various categories of narrative criteria TMDLs. Types of impairment documented in the review consisted of toxicity, toxic pollutants, excess sedimentation, nutrients, and other (e.g., debris, sludge deposits).

The second topic of the preliminary review concerned the inclusion of waterbodies on the 303(d) list because of narrative criteria impairments. The National Research Council review of the TMDL process (NRC, 2001), and the WERF project "Navigating the TMDL Process: Listing and Delisting" (Thornton and Freedman, 2002), have recommended changes to the 303(d) listing process. These changes include dividing the 303(d) list into two parts: 1) a "preliminary" list of waterbodies that require further evaluation prior to conducting a TMDL, and 2) a TMDL "action" list. Thornton and Freedman (2002) recommended that all waterbodies listed as impaired because of nonattainment of narrative criteria be placed on the preliminary list, because of the difficulty of defining impairment in the absence of quantitative criteria. These waterbodies would not be moved to the action list until numeric criteria were developed. To address this issue, one question of the preliminary review was "Is the waterbody obviously impaired?" While this question was subjective, it helped characterize the TMDLs with respect to listing issues, and allowed for further discussion of the types of impairments in each category. In practice, TMDLs that listed any observed impairment of the designated use were categorized as impaired, while those that provided no justification of impairment, or based strictly on evaluated data, were categorized as "not obviously impaired."

The final topic of the preliminary review involved the determination of a target measure and level for the TMDL analysis. A primary concern for narrative criteria TMDLs is the selection of an objective quantitative measure to represent a narrative criterion, rather than over-reliance on subjective judgment. The survey investigated the use of judgment decisions in criteria selection using two questions:

1. **Was the target measure a direct measure of the narrative criterion or a surrogate measure?** In some cases, the TMDL target is defined for a surrogate measure instead of the pollutant responsible for the impairment (e.g., bank stability as a surrogate for excess sedimentation). This question was intended to define whether a judgment decision was required in selecting the surrogate measure to be used for the TMDL target. For this question, the TMDLs were divided into the categories of Direct Measure (i.e., the target measure is the same as the pollutant) or Surrogate (i.e., a surrogate measure was selected).
2. **How was the target level selected?** This question was intended to define whether a judgment decision was required in selecting the target level. Results were divided into one of three categories:
   - Criteria: An existing federal criterion/state standard or numeric translator was used;
   - Literature: The target level was taken from the scientific literature; or
   - Best Professional Judgment (BPJ): Best professional judgment on the part of the TMDL developer was used.

A summary of the preliminary review findings is presented in Table 2-1.

Table 2-1. Results of Preliminary Review of Narrative Criteria TMDLs.

Type of Narrative Criteria	Total	Obviously Impaired?		Target Measure[1]		Target Level Selection[2]			
		Yes	No	Direct Measure	Surrogate	Criteria	Literature	BPJ	Not certain
Toxicity	0	-	-	-	-	-	-	-	-
Toxic pollutant									
Human health	4	4	-	4	-	4	-	-	-
Aquatic life	1	1	-	1	-	1	-	-	-
Sedimentation	10	7	3	8	5	2	2	8	-
Nutrients	10	8	2	7	6	1	3	8	1
Other	3	1	2	2	1	2	-	-	1
Total	28	21	7	22	12	10	5	16	2

Notes: 1) Some TMDLs used both a direct measure and a surrogate.
2) Some TMDLs used a combination of methods for target level selection.

The preliminary review findings provided valuable information for use in compiling additional narrative TMDLs and structuring a detailed review. The preliminary review identified the need to expand the review beyond approved TMDLs to include draft TMDLs. This was needed to allow for a review of narrative TMDL categories that presently have a limited number of approved TMDLs, such as toxicity. The preliminary review also confirmed the areas of suspected concern when addressing narrative criteria in the TMDL process, as discussed in Section 1.3, and provided specific examples of the challenges. These examples were then used to structure and refine detailed review questions.

2.2.2 Compilation of Additional Narrative Criteria TMDLs

The database of TMDLs from the preliminary review was augmented with additional TMDLs from a variety of sources. The intent of the compilation was to increase the breadth and depth of TMDLs to be reviewed. Additional narrative criteria TMDLs were identified by:

- Collection of additional narrative TMDLs from the WERF database;
- Outreach to U.S. EPA Headquarters;
- Outreach to U.S. EPA Regional TMDL coordinators;
- Outreach to state TMDL coordinators; and
- Searches on the Internet.

The preliminary review of TMDLs considered only the narrative criteria subset of the 176 TMDLs reviewed for the WERF project "Critical Assessment of the TMDL Process: Evaluation and Improvement" (Limno-Tech et al., 2003). The WERF project, however, had compiled a larger database containing 212 TMDLs; for the current project, the entire database of 212 TMDLs was considered, and additional narrative criteria TMDLs were obtained for review.

Individuals at U.S. EPA Headquarters involved in TMDL development were contacted to inquire about the availability of narrative TMDLs for specific impairment types. This outreach yielded approximately 60 additional TMDLs. Special effort was made to obtain draft or final TMDLs for waterbodies impaired due to toxicity; the project team contacted regional and state TMDL coordinators and used U.S. EPA on-line databases, but no additional toxicity TMDLs were located.

The final result of these efforts to obtain additional narrative TMDLs was the collection of over 120 narrative TMDLs.

One objective of the compilation process was to obtain a database of narrative TMDLs with a distribution of impairment types consistent with the distribution of impairment types found on 303(d) lists nation-wide. This was achieved by: 1) defining the national distribution of all types of impairment; 2) refining this distribution to reflect impairments typically associated with narrative criteria; and 3) reviewing the database of narrative TMDLs obtained to find a subset that contained the desired distribution.

The distribution of 303(d) listed waterbodies categorized by type of impairment was obtained directly from U.S. EPA's list of "Top 100 Impairments" available on U.S. EPA's TMDL website (http://www.epa.gov/owow/tmdl, December 2001). This distribution was available only in combined form for both numeric and narrative criteria. Therefore, each impairment type was classified as either numeric or narrative, depending upon the criterion most commonly associated with the impairment. Some impairment types were grouped when determined to be similar. Refining the distribution of impairment types on the national 303(d) list to reflect only narrative criteria resulted in the distribution shown in Table 2-2. The impairments most commonly associated with narrative criteria were excess sedimentation, nutrients, habitat alterations, toxicity, and biology.

Table 2-2. Distribution of Impairment Types on the National 303(d) List Most Commonly Associated with Narrative Criteria.

Impairment Type	Number of Occurrences	% of Occurrences
Excess Sedimentation	6528	34.5%
Nutrients	4749	25.1%
Other Habitat Alterations	3208	17.0%
Toxicity	1555	8.2%
Biology	1236	6.5%
Miscellaneous	836	4.4%
Noxious Plants	800	4.2%

The subset of TMDLs selected for final review was intended to match the distribution of impairment types consistent with that shown in Table 2-2. Initial review of the TMDLs showed, however, that an exact comparison could not be achieved. The review showed inconsistencies in how waterbodies are placed in the U.S. EPA categories, especially when the pollutant/stressor is different from the water quality endpoint. For example, excess nutrient loading leading to violation of the dissolved oxygen standard was categorized either as being impaired by nutrients or dissolved oxygen. Excess sedimentation leading to habitat impairment could be classified as being impaired by sediment or habitat (or several other categories). For this reason, certain impairment categories were consolidated, and only a qualitative comparison was conducted to ensure that the distribution of impairment types reviewed was consistent with the distribution of impairment types observed on the national 303(d) list. In addition, some impairment types (most notably ambient toxicity) were under-represented in the TMDLs available for review, despite outreach efforts to obtain draft or final toxicity TMDLs. The final distribution of the 120 TMDLs selected for review is shown in Table 2-3.

Table 2-3. Distribution of Impairment Types for Narrative Criteria TMDLs Selected for Review.

Impairment Type	Number of Occurrences	% of Occurrences
Excess Sedimentation/Habitat	48	40.0
Nutrients/Noxious Plants	39	32.5
Toxicity/Biology	23	19.2
Miscellaneous/Unclear	10	8.3

The geographic distribution of the compiled narrative TMDLs was broad (Figure 2-1). The only area underrepresented was the Northeast (U.S. EPA Regions I and II), although this is partially compensated by the large number of TMDLs in nearby Region III.

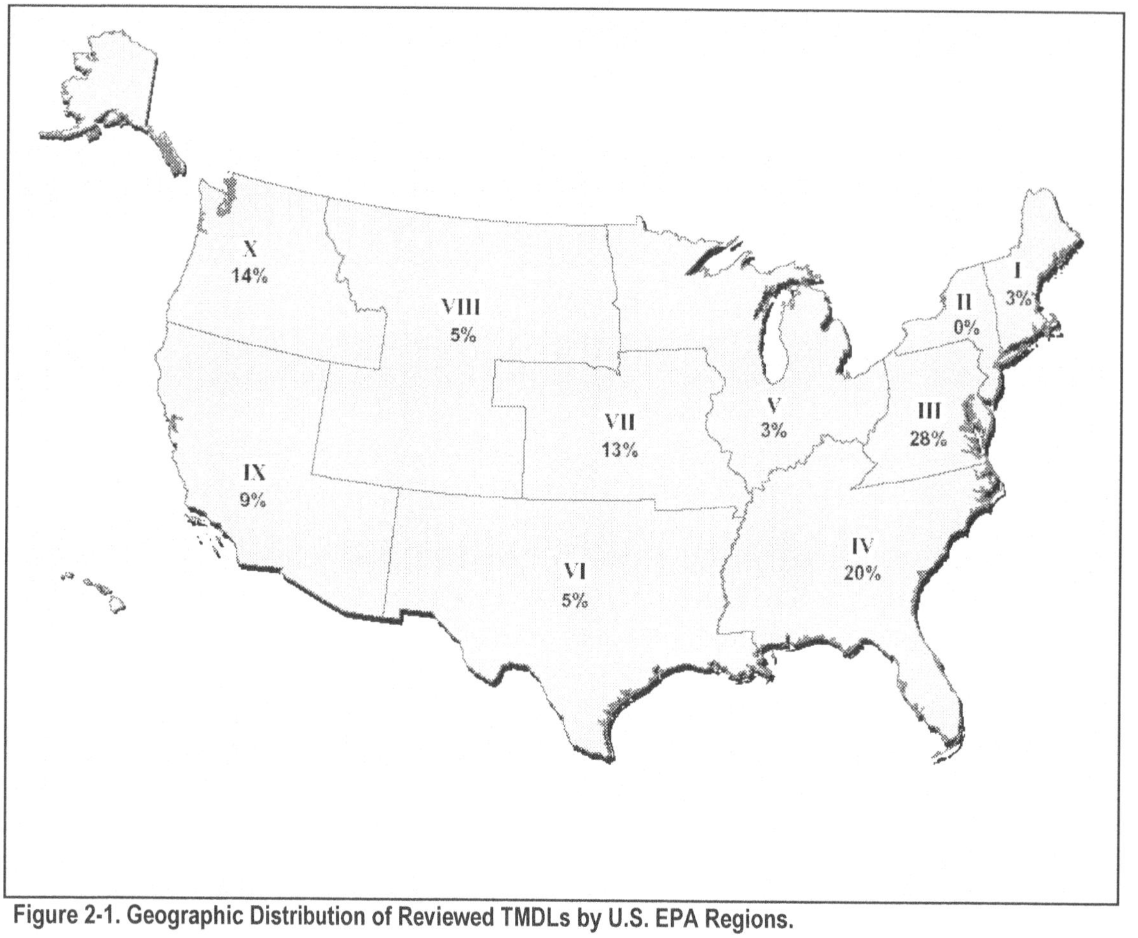

Figure 2-1. Geographic Distribution of Reviewed TMDLs by U.S. EPA Regions.

The geographic distribution of the impairment types also was reviewed. Regional water quality issues influence the geographic distribution of impairment types. Each region may place emphasis on particular pollutants or impairment types. For example, the impact of excess sedimentation on salmonid spawning and rearing receives specific attention in the Northwest (in U.S. EPA Region 10). Individual states or U.S. EPA regions also may emphasize a particular pollutant or impairment type based not only on regional water quality issues, but also because of the form of the state water quality standards or specific measurement techniques. For example, Ohio's promulgated biocriteria include two indices based on stream fish assemblages (Index of Biotic Integrity, IBI, and the Modified Index of Well-Being, MIwb) and one index based on stream macroinvertebrate assemblages (Invertebrate Community Index, ICI). As another example, U.S. EPA Region IV has had a historical emphasis on the use of whole effluent toxicity testing. The majority (55%) of the excess sedimentation TMDLs were from the three western U.S. EPA Regions (VII, IX, and X), while 89% of the reviewed nutrient TMDLs were from the Midwest and East. The majority of toxicity TMDLs came from Region IV. U.S. EPA. TMDL coordinators from across the country were contacted to obtain a broader geographic distribution of toxicity TMDLs, but no additional TMDLs were identified. The other types of impairments showed no strong geographic trend. The geographic distribution by impairment type for the reviewed TMDLs was not dissimilar to the geographic distribution observed for all 303(d)-listed waterbodies, and the database of TMDLs was deemed representative for review purposes.

2.2.3 Structure of Narrative TMDL Review

The full set of 120 narrative TMDLs was reviewed to characterize the current "state of the practice" of TMDL development for narrative criteria. The review of each TMDL included four topic areas:

1. Defining impairment;
2. Determining the causes of impairment;
3. Setting TMDL target measures and levels; and
4. Other noteworthy features.

2.2.3.1 Defining Impairment

The first review topic focused on the reasons that the waterbody was listed as impaired and was also designed to ensure that a representative subset of narrative criteria TMDLs were being reviewed. This review topic also provided insight into how waterbodies were listed for narrative criteria and whether the listing process needs refinement. The first series of review questions served strictly to characterize the reasons for listing:

- U.S. EPA impairment type;
- Pollutant/stressor responsible for listing; and
- Water quality measure responsible for listing.

The questions were designed to characterize the TMDLs to allow for the comparison of the distribution of impairment types with the types seen on 303(d) lists nationwide. The latter two questions were designed to provide more specificity on the distribution of pollutants/stressors to be controlled, as well as the water quality measures.

The final review question in this topic area was designed to provide insight into the certainty that the waterbody was impaired because of nonattainment of a narrative criterion. Each 303(d) listed waterbody was evaluated on whether there was obvious impairment, and then it was placed into one of two categories:

- Listed because of an indirect indicator or surrogate measure; or
- Listed because of direct impairment of the use.

An example of the first category would be a waterbody listed as having impaired biological use based upon an observed suspended solids concentration above a certain threshold. An example of the second category would be a waterbody listed as having impaired biological use based upon the absence of fish species expected to occur at the site. Waterbodies in the first category have the greatest potential for being falsely listed, depending upon the site-specific understanding between the surrogate measure and the narrative criterion (discussed subsequently). Also, it should not be assumed that all impairment determinations based on direct measures of impairment are correct. Some sites have been improperly listed by comparing a direct measure to an inappropriate threshold (e.g., the absence of salinity-intolerant fish species in a river downstream of naturally occurring salt springs). The review did not presuppose that all waterbodies listed based on a surrogate measure were in error; rather, the review was attempting to characterize what percentage of waterbodies fall in this category and, therefore, have a greater likelihood of error.

2.2.3.2 Determining Causes of Impairment

The second review topic focused on the understanding of site-specific cause-and-effect mechanisms and served multiple purposes. This topic assessed the potential for inappropriate identification of the causes of impairment, and the possibility that the TMDL was developed to control the wrong pollutant. For those TMDLs listed because of a surrogate measure, this topic provided information on whether there was sufficient knowledge to define the waterbody as impaired. In addition, each TMDL was initially placed into one of two categories:

- The linkage between causes and impairment was implicitly assumed; or
- The linkage between causes and impairment was clearly documented.

An example of misidentification of causes in the first category would be the assumption that an observed lack of biodiversity was caused by effluent toxicity, without considering any habitat impacts.

2.2.3.3 Setting TMDL Targets

The third review topic focused on how the final TMDL target was selected and was intended to provide insight on the degree of judgment required on the part of the TMDL developer. Each TMDL reviewed was placed into one of the following categories on the basis of the setting of the TMDL target:

- Existing federal criterion or numeric translator that had not been formally adopted as a state standard;
- Literature; or
- Best professional judgment.

For those waterbodies where the TMDL target level was based upon literature and/or best professional judgment, the review documented the specific assumptions made in defining the target level.

2.2.3.4 Other Significant Features

The final review topic was designed to capture any significant features of narrative TMDL development not captured by the earlier topics. These features could consist of items such as:

- Aspects of narrative TMDL development done particularly well;
- Aspects of narrative TMDL development done particularly poorly; and
- Unique challenges of narrative TMDL development not captured previously.

The results of this review are presented in the sections that follow, grouped first by the category of narrative criteria (toxicological, ecological, and aesthetic), and grouped within each of those by step in the TMDL process (defining impairment, determining the causes of impairment, and setting TMDL target measures and levels). A listing of the TMDLs reviewed is included in Appendix A. The guiding principles that were developed in response to these findings are presented in Chapters 3.0 - 5.0, and summarized in Appendix B.

2.3 Summary of Review Findings

The review identified problem areas for each of the three categories of narrative criteria and in each of the three steps of the TMDL process that were evaluated. A summary of the results of the review is presented in Table 2-4 and discussed briefly below. The remainder of this chapter presents detailed findings of the review for each of the three categories of narrative criteria, and more specifically for each of the three steps in the TMDL process presenting challenges unique to narrative criteria. Chapters 3.0-5.0 of this report present guiding principles for each of the three steps in the TMDL process, specifically responding to these findings.

Table 2-4. Summary of Challenges Identified in the Review of Narrative Criteria TMDLs.

Categories of Narrative Criteria	Steps in the TMDL Process Presenting Challenges Unique to Narrative Criteria		
	Defining Impairment	**Determining Causes of Impairment**	**Setting TMDL Target Measures and Levels**
Toxicological	Use of evaluated data to presume impairment Interpreting ambient toxicity data; and Listing based on fish consumption advisory	Determining cause based on ambient or effluent toxicity data	Setting fish consumption TMDL targets in water column for complex chemical speciation and uptake
Ecological	Use of surrogate measures may not be reflective of criteria Selection of impairment endpoint inappropriate for the waterbody	Complex and poorly understood cause-and-effect mechanisms	Selecting target measures reflective of criteria and also able to be linked to controls
Aesthetic	Converting perceptions to levels	---	Converting perceptions to levels

For the TMDLs reviewed, toxicological TMDLs were conducted primarily because of the results of aquatic toxicity tests, and human health concerns caused by fish tissue contamination. When defining impairment, toxicity TMDLs were identified as posing problems when using evaluated data to presume impairment and also when interpreting ambient toxicity data. Challenges were also observed when a waterbody was listed as impaired strictly because of the presence of fish consumption advisories; the presence of consumption advisories does not necessarily indicate impairment (U.S. EPA, 2000b). Determining the causes or sources of impairment based on ambient toxicity data was also found to be a major challenge because of the lack of data describing the cause of toxicity (e.g., toxicity identification evaluation). Finally, the results of the review indicated that it is difficult to set TMDL target levels in the water column that will result in acceptable fish tissue concentrations for some chemicals, such as mercury, that undergo complex speciation in the natural environment and uptake through the food chain.

Reviewed TMDLs in the Ecological category consisted primarily of excessive sedimentation/habitat impairments, with some eutrophication TMDLs also falling into this category. Impairment often was defined using surrogate measures that were poorly correlated to the designated use for this category of narrative TMDLs. While all criteria are "surrogate measures" of the designated use, some are much more closely correlated to the use than others. For example, the amount of land in agricultural production in a watershed was used as a surrogate indicator of pesticide toxicity; direct measures of ambient toxicity would be much more closely correlated to the designated use. The review also noted that impairment decisions were sometimes based on an inappropriate target level for the waterbody. For example, the target level of a macrobenthos index for a channelized urban stream should reflect the local habitat quality, and would be different from the target level for a pristine watershed. Determining the causes of impairment presented significant challenges for TMDLs in this category because of complex and typically poorly understood cause-and-effect relationships between stressors and ecological responses. Finally, selecting TMDL target measures that were both reflective of the narrative criterion and were linked to TMDL controls proved difficult for this category. In practice, TMDLs that listed any observed impairment of the designated use were categorized by the project team as impaired, while those that provided no justification of impairment, or based strictly on evaluated data, were categorized as "not obviously impaired."

Finally, TMDLs in the Aesthetic category were developed for waterbodies impaired by turbidity, debris, or the consequences of eutrophication. The primary challenge in these TMDLs was determining appropriate levels for defining impairment and setting TMDL targets (i.e., the quantity of a given constituent/characteristic that constitutes a nuisance level). This process requires converting human perceptions into levels for aesthetic impairments.

2.4 Review Findings for Toxicological Impairments

A total of 23 of the reviewed TMDLs corresponded to toxicological impairments (ambient aquatic toxicity and fish tissue contamination). As discussed below, the primary areas requiring guiding principles for this category of narrative criteria are in the determination of impairment from ambient toxicity or fish consumption advisory, and in determining the ultimate causes and sources of toxicity. Finally, the results of the review indicated that it is difficult to set TMDL target levels in the water column that will result in acceptable fish tissue concentrations for some chemicals, such as mercury, that undergo complex speciation in the natural environment and uptake through the food chain. Furthermore, air deposition of mercury can increase the complexity of addressing appropriate controls and load distribution to local point sources.

The narrative criteria associated with toxic impairments generally take the form of requiring "no toxics in toxic amounts." A sampling of the specific narrative criteria language observed in the reviewed TMDLs is listed below:

- "all waters shall be free from toxic, corrosive, acidic and caustic substances discharged from municipalities, industries or other sources, such as nonpoint sources, in amounts, concentrations or combinations which are harmful to humans, animals or aquatic life."
- "there may be no concentrations or toxics substances in water or in shoreline or bottom sediments, that, singly or in combination, cause, or reasonably can be expected to cause, toxic effects on aquatic life."
- "As a result of man-caused point or nonpoint source discharge, waters of the State must not contain: Hazardous materials ... in concentrations found to be of public health significance or to adversely affect designated or protected beneficial uses."
- "Toxic substances shall not be introduced above natural background levels in waters of the state which may adversely affect characteristic water uses, cause acute or chronic conditions to the aquatic biota, or adversely affect public health."

- "Waters shall be free from substances attributable to wastewater discharges or agricultural practices in concentrations or combinations, which are acutely toxic to humans, animals or plant life."

Toxicological TMDLs were conducted primarily for two types of impairments: 1) the presence of ambient toxicity to aquatic life; and 2) human health concerns from fish tissue contamination. Of the 23 toxicological TMDLs reviewed, 12 related to ambient aquatic toxicity and 11 related to human health concerns from fish tissue contamination.

2.4.1 Defining Toxicological Impairment

Determination of impairment was found to be a concern for aquatic toxicity TMDLs as well as for fish tissue contamination. The results of the review show weak bases for listing decisions. Specific issues were identified related to basing aquatic toxicity impairment decisions solely on land use in the watershed, and tissue contamination decisions solely on the presence of consumption advisories.

The majority (92%) of the aquatic toxicity TMDLs reviewed was based on evaluated data that provided an indirect and potentially incorrect measure of impairment. The primary reason for listing was the presence of whole effluent toxicity in the effluent of point source discharges to effluent-dominated streams. The rationale behind these listing decisions was that these streams consisted almost completely of effluent during certain periods of the year, and that the presence of toxicity in the effluent would directly translate to instream toxicity during low flow periods. This assumption is consistent with the findings of U.S. EPA (1991) Complex Effluent Toxicity Testing Program studies, which found that impairment of in-stream uses was correlated to the presence of whole effluent toxicity in discharges. Recent research (Diamond et al., 1999), using a more extensive and recent data set, drew the following conclusion regarding the correlation between observed whole effluent toxicity and stream impairment:

- The frequency of occurrence of WET test failures was highly correlated with the presence of stream impairment; dischargers failing at least 25% of their WET tests exhibited a higher probability of impairment than dischargers with fewer WET failures.
- Dischargers that conducted fewer than three WET tests in the year prior to collection of bioassessment data had especially poor relationships between WET test results and stream biological conditions.
- Effluent dominated systems had a higher probability of being associated with stream impairment if the discharger failed several WET tests.

Several other aquatic toxicity TMDLs based the determination of impairment solely on the presence of agricultural land uses, which were presumed to result in ambient toxicity.

The use of evaluated data as a basis for 303(d) listing, has been a source of controversy. The NRC (2001) has suggested placing these types of waterbodies on a preliminary list that requires priority monitoring to determine the need for TMDL development. For the TMDLs reviewed as part of this research project, the presence of whole effluent toxicity in the effluent of point source discharges to effluent-dominated streams shows the potential for impairment. Land-use data, without supporting chemical or biological data from the waterbody, do not provide a definitive basis for listing a waterbody as impaired.

Only one of the 12 reviewed aquatic toxicity TMDLs based the impairment decision on ambient toxicity testing data, and even here the rationale used for concluding that impairment existed was not provided. A review of the toxicity data provided showed that improper consideration of variability in the toxicity test method might have led to a faulty impairment determination. (This issue is discussed in more detail in Chapter 3.0 in a case example demonstrating one of the guiding principles.) The lack of available ambient toxicity TMDLs for review relative to the large number of 303(d) listings indicated that the need exists for guiding principles for determination of impairment for aquatic toxicity.

Determination of impairment was straightforward for fish tissue contamination for those TMDLs that were reviewed under this project, as numeric targets representing appropriate human health action levels were available for all of the reviewed TMDLs. However, specific issues were identified as part of a broader review of 303(d) lists nationwide. These issues resulted from basing the determination of impairment solely on the presence of fish consumption advisories, even for waters where no site-specific data were available to indicate impairment. For example, all of Michigan's inland lakes were included on the 1998 303(d) list because of a generic mercury fish consumption advisory (U.S. EPA, 1998a), even though mercury data were available only for a subset of Michigan lakes. This is contrary to present U.S. EPA guidance (2000b) that a waterbody should not be listed as impaired strictly because of the presence of a fish consumption advisory, unless the advisory is based upon site-specific data that demonstrates that the human health risks exceed those listed in the state water quality standards. Also, some fish consumption advisories are based on criteria (e.g., proximity to a discharge) that are not suitable for an impairment decision (U.S. EPA, 2000b). Guiding principles will be provided in Chapter 3.0 relative to determining impairment because of fish tissue contamination.

2.4.2 Determining Causes of Toxicological Impairment

Specification of cause-and-effect linkages was found to be a potential concern for both aquatic toxicity and fish tissue contamination TMDLs. The review findings showed that the cause-and-effect linkage was not considered for those toxicity TMDLs where impairment was caused by a single discharge of whole effluent toxicity to an effluent dominated stream, as no ambient impairment was ever measured. Also, the review revealed that no attempt was made at these sites to identify the specific contaminants responsible for the toxicity. The one TMDL that considered multiple sources of toxicity was unable to resolve the individual contributions of toxicity from the contributing sources. That TMDL assumed that the chronic toxicity from a wide range of sources and different test species could be treated in an additive fashion. This is contrary to current scientific understanding (e.g., Rand, 1995). Although only a single TMDL was reviewed that considered multiple sources of toxicity, the observed issues may be indicative of other similar waterbodies on 303(d) lists for aquatic toxicity impairment. Because of the potential significance of this issue, guiding principles should be developed for determining the cause of ambient toxicity when the potential for multiple sources exists.

Relatively few issues were discovered in the review associated with determining the causes of impairment for fish tissue contamination TMDLs. Little uncertainty existed in defining the specific sources of the problem, because a single, well-identified contaminant was responsible for the impairment. However, a broader review indicated the potential for challenging issues. First, certain chemicals (such as mercury) undergo complex speciation reactions that affect bioavailability. For chemicals such as mercury, the source can play a direct role in its ultimate bioavailability, and therefore complicate the identification of the true causes of impairment. The guiding principles presented in Chapter 4.0 for determining the causes of impairment apply in these situations.

2.4.3 Setting Toxicological TMDL Targets

Selection of a TMDL target was relatively straightforward in the TMDLs reviewed, both for aquatic toxicity and fish tissue contamination. All of the aquatic toxicity TMDLs used a TMDL target of 1.0 unit of chronic toxicity (TU_c), which is derived by dividing 100 by the No Observed Effect Concentration (NOEC) (U.S. EPA, 1991). As mentioned above, human health target levels were readily available on a state-specific basis for fish tissue contamination for most sites and little additional judgment was required on the part of the TMDL developer. Again, mercury was found to be the exception, as selection of an appropriate water column target for mercury is dependent on the site-specific speciation of mercury. The primary source of uncertainty in this type of TMDL was defining the quantitative linkage between water column (and sediment) concentrations and fish tissue levels. Although this is a challenging scientific issue, this uncertainty is not unique to narrative criteria and is not specifically addressed by this research.

2.5 Review Findings for Ecological Impairments

A total of 48 of the narrative criteria TMDLs reviewed corresponded to ecological impairments, consisting primarily of excess sedimentation/habitat impairments and nutrient impairments. As discussed below, the primary challenges in these TMDLs were complex and poorly understood cause-and-effect relationships between stressors and ecological responses. Two-thirds of these TMDLs listed the cause of the impairment as known, without any explicit justification. Specific challenges included determining the causes responsible for the observed impairment, and selecting appropriate TMDL target measures that were both reflective of the narrative criterion and able to be linked to TMDL controls.

The narrative criteria associated with ecological impairments generally take the form of "maintaining the biological integrity of the water." A sampling of the specific narrative criteria language observed in the reviewed TMDLs is listed below:

- "maintain the biological integrity of the waters of the State."
- "no significant adverse impact to the chemical, physical, hydrological, or biological components of aquatic ecosystems shall be allowed."
- "Suspended solids added to surface waters by artificial sources shall not interfere with the behavior, reproduction, physical habitat or other factor related to the survival and propagation of aquatic or semi-aquatic or terrestrial wildlife."
- "Any activity causing water pollution that would significantly degrade the biological integrity of a body of water or significantly impact or displace an identified key species shall not be allowed."
- "The introduction of plant nutrients into streams, lakes or wetlands from artificial sources shall be controlled to prevent the accelerated succession or replacement of aquatic biota."
- "Surface waters shall be free from pollutants in amounts or combinations that…cause the growth of algae or aquatic plants that inhibit or prohibit the habitation, growth, or propagation of other aquatic life …"

A wide range of impairment types was found in the Ecological Impairments category. The most common causes were excess sediment and nutrient loads. The most common impairment endpoints were habitat, macrobenthos, fish community and chlorophyll *a* concentration. Eutrophication impairments could be considered under either the Ecological

Impairments or Aesthetic Impairments category, depending upon the nature of the impairment. This section considers only those cases where excess nutrient loads were viewed as impairment to ecological health; eutrophication impairments of recreational use were considered in the Aesthetic Impairments section.

2.5.1 Defining Ecological Impairment

A wide range of methods was used in the reviewed TMDLs to determine the existence of ecological impairment. The majority of waterbodies based the impairment decision on direct observation of the impairment based on the narrative criterion, although 10% of the TMDLs based the impairment decision on data for a surrogate measure intended to reflect the narrative criterion. Indicators used to define ecological impairment in the reviewed TMDLs are listed below, categorized by whether the impairment decision was based upon a direct measure of the narrative criterion or a surrogate measure.

- Impaired uses:
  - Reduced fish populations compared to historical levels;
  - Impaired fish spawning;
  - Low macroinvertebrate index; and
  - Presence of algal blooms.
- Surrogate measures of impaired uses:
  - Sedimentation rate viewed as being in excess of the level necessary to protect the fishery;
  - Turbidity levels as an indicator of salmon displacement;
  - Impaired habitat indicator;
  - Observed bottom deposits presumed to affect fish and shellfish propagation; and
  - Chlorophyll *a* concentrations.

The primary concern for these TMDLs was how well the surrogate measures reflected the narrative criterion. The use of a surrogate measure that poorly reflects the criterion could result in the 303(d) listing of a waterbody that is not impaired, or the nonlisting of a waterbody that is impaired. The root of this problem is that the cause-and-effect relationship between stressors and ecological response is often poorly understood, as discussed below.

2.5.2 Determining Causes of Ecological Impairment

An understanding of the cause-and-effect linkage plays a key role in two aspects of the TMDL process for ecological impairments, as the complex stressor-response relationships can make determination of the causes of the impairment difficult. An understanding of the causative stressors is essential to focus TMDL control efforts. Specification of cause-and-effect linkages was found to be a potential concern for both the fish and benthos TMDLs, but of less concern for nutrient issues. As discussed below, the need for guiding principles in developing the cause-and-effect linkages is crucial.

For the majority of fish and benthos impairments, excess sediment load was identified as the stressor of concern. Only one-third of the TMDLs reviewed conducted an explicit assessment of the cause of impairment, either in terms of documenting that excess sedimentation was the primary stressor or defining the primary source of the excess sediment. The remaining two-thirds of these TMDLs listed the cause of impairment as known, without any explicit justification. The remaining non-nutrient ecological impairment TMDLs also lacked explicit documentation of the presumed cause-effect relationship. This absence of documentation (and potentially understanding) of the stressor-response relationship in the majority of ecological impairment TMDLs calls for the development of guiding principles.

For ecological impairment from nutrient enrichment, the relevant cause-and-effect mechanisms between nutrient loading and plant response were well identified. The majority of these TMDLs provided explicit demonstration of the limiting nutrients, as well as an assessment of the source of nutrients throughout the watershed. The quantitative relationship assumed between nutrient loads and ecological response was sometimes poorly documented, but poor documentation was not unique to narrative criteria, and did not require the development of additional guiding principles as part of this research.

2.5.3 Setting Ecological TMDL Targets

A wide variety of TMDL target measures were used for the ecological TMDLs. For the fish and benthos TMDLs, target measures consisted of:

- Allowable watershed solids load
- Maximum allowable TSS concentration
- Substrate composition
- Road condition indicators
- Hillslope condition indicators
- Turbidity
- Pool frequency/depth
- Abundance of large woody debris
- Areal extent of suitable habitat
- Frequency of future dredging
- Fish population
- Fish age distribution
- Channel form/structure
- Macroinvertebrate indices
- Streambank stability

Best professional judgment was the most commonly used method for selection of target levels. Best professional judgment is subjective and can lead to inconsistency in setting TMDL targets. Also, many of the analysts developing the TMDL may lack the expertise to apply best professional judgment. Data from reference sites and the scientific literature also were used to select target levels. Obstacles are encountered when using reference sites for habitat-limited areas where use of a traditional unimpacted or pristine reference site may not be appropriate (Section 5.4.1.2 discusses reference watersheds in greater detail). Scientific literature values are not always the best solution either, as they lack site specificity.

The most significant issues found in the selection of these targets were: 1) many target levels were selected through the use of best professional judgment, often with the absence of site-specific data necessary to allow an informed judgment; 2) many of the targets were for stressors and contained a high degree of uncertainty regarding whether achievement of these targets would lead to attainment of the narrative criterion; and 3) many targets reflective of the narrative criterion could not be quantitatively linked back to necessary TMDL controls. Development of guiding principles for selection of TMDL targets for ecological impairments was therefore recommended.

For the nutrient TMDLs, targets primarily consisted of chlorophyll *a* levels, nutrient concentrations, or some combination of the above (along with water transparency) reflected in a trophic status index (TSI). Literature values, derived from sources such as the U.S. EPA National Eutrophication Survey (1974), were the most common source of the target levels. Given recent U.S. EPA (1999b) guidance in TMDL development for nutrients, along with efforts to define water quality criteria for nutrients (U.S. EPA, 2002a; Grubbs, 2001), the development of specific guiding principles unique to eutrophication targets is not seen as a priority for this research. However, the guiding principles presented in Chapter 5.0 (such as ensuring that impairment targets are directly linked to the site-specific designated use) are applicable to nutrient criteria. Specifically, careful attention needs to be paid to ensuring that there is a clear linkage between uses being protected and the nutrient or surrogate thresholds that are being considered as targets.

2.6 Review Findings for Aesthetic Impairments

A total of 49 of the reviewed TMDLs corresponded to aesthetic impairment, consisting primarily of debris, turbidity, and nutrient impairment. As discussed below, the primary challenge encountered in these TMDLs was defining impairment and setting appropriate TMDL targets; that is, determining what quantity of a given constituent constitutes a nuisance level. None of the TMDLs reviewed included a rigorous assessment to define appropriate levels for defining impairment. Guiding principles for converting perceptions into levels for aesthetic impairments were therefore researched.

The narrative criteria associated with aesthetic impairments generally take the form of "waters will contain no nuisance amounts of ...." A sampling of the specific narrative criterion language observed in the reviewed TMDLs is listed below:

- "Waters shall be free from substances in sufficient amounts to cause unsightly color or turbidity, offensive odor or prevent full maintenance of beneficial use."
- "Nuisance growths are prohibited."
- "Aesthetic values shall not be impaired by the presence of materials or their effects, excluding those of natural origin, which offend the senses of sight, smell, touch, or taste."
- "waters shall ... be free of culturally induced algae blooms which impair their potential use and enjoyment."
- "prohibit pollution of waters of the State by any material in amounts sufficient to create a nuisance."
- "Waters shall not contain suspended material in concentrations that cause a nuisance."
- "Waters shall be free from substances in sufficient amounts to cause the formation of putrescent, unsightly or harmful bottom deposits or prevent full maintenance of beneficial uses."
- "The introduction of plant nutrients into surface waters designated for primary or secondary contact recreation use shall be controlled to prevent the development of objectional concentrations of algae or algal byproducts or nuisance growths of submersed, floating, or emergent aquatic vegetation."
- "Surface waters shall be free from pollutants in amounts or combinations that ...cause the growth of algae or aquatic plants that ... impair recreational uses ..."

- "Waters shall be free from materials attributable to wastewater discharges or agricultural practices producing objectionable color, odor or other aesthetically objectionable conditions."

A wide range of impairment types was found relating to aesthetic criteria. The most common reasons given for listing waterbodies as impaired were excess debris, nuisance levels of plant growth caused by excess nutrient loading, reduced reservoir capacity, and turbidity.

2.6.1 Defining Aesthetic Impairment

The determination of impairment was not straightforward for the aesthetic category. While essentially all of the TMDLs reviewed used a direct measure of the designated use as opposed to some surrogate measure, determining the level that constitutes impairment was subject to wide interpretation on the part of the TMDL developers. TMDLs for impairment by debris were based on observations of debris; turbidity TMDLs were based on turbidity measurements, and so on. But determining what quantity of a given constituent constitutes a nuisance level is complex. Aesthetic considerations can be site-specific and vary widely from region to region because of the subjective nature of aesthetics. None of the TMDLs reviewed included a rigorous assessment to define thresholds for aesthetic impairment. The debris TMDLs consistently used a level of zero debris. Eutrophication TMDLs used national literature values, even in cases where they were not appropriate for site-specific conditions. Best professional judgment (in the absence of site-specific data) was used in setting levels for many of the other aesthetic TMDLs. Guiding principles were therefore recommended for converting perceptions into levels for aesthetic impairments.

2.6.2 Determining Causes of Aesthetic Impairment

Most of the TMDLs for aesthetic impairments had a clearly defined cause-effect linkage between a single source and the resulting endpoint. The primary uncertainty was in quantitatively defining the relationship between loading and the resulting ambient concentration, such as for eutrophication. This uncertainty is not unique to narrative criteria and no additional guiding principles are recommended for this area as part of this research.

2.6.3 Setting Aesthetic TMDL Targets

Determining appropriate TMDL target levels in the aesthetic impairment TMDLs raises essentially the same issues as defining impairment thresholds. This is because almost all of the TMDLs reviewed used a direct measure of the narrative criterion in the determination of impairment and also as the TMDL target. The primary uncertainty with this is quantitatively defining the relationship between a proposed source control and the resulting ambient concentration. This uncertainty is not unique to narrative criteria and no additional guiding principles are recommended for this area.

CHAPTER 3.0

GUIDING PRINCIPLES FOR DEFINING IMPAIRMENT

3.1 Introduction

The TMDL process begins with a determination of impairment by the state agency, identifying waterbodies failing to support their designated uses and placing them on the state's 303(d) list. This is an important step and errors made here can result in either of two undesirable consequences. The first type of error is listing an un-impaired waterbody as impaired. This can lead to wasted expenditures in developing TMDLs and implementing control measures when none are necessary, as well as significant permitting ramifications for permittees currently discharging to the system. The second type of error is not listing impaired waterbodies, which results in inadequate protection of the nation's waters. Defining impairment by nonattainment of narrative criteria, which lack specific numeric measures, is particularly susceptible to errors.

This chapter contains basic guiding principles to improve the determination of impairment based on narrative criteria. It begins with a discussion of the challenges and concerns associated with defining impairment based on narrative criteria. These challenges and concerns were identified in the review of narrative TMDLs. Guiding principles are then presented, categorized by the challenges and concerns they are intended to remedy. This chapter is divided into sections:

Challenges and concerns (Section 3.2)
Consistent decision process (Section 3.3)
Selection of measures and levels of impairment (Section 3.4)
Data interpretation (Section 3.5)
Determining aesthetic impairment (Section 3.6)

3.2 Challenges and Concerns

The TMDL process begins with the definition of waterbodies failing to support their designated uses. The determination of impairment based on narrative criteria can be particularly challenging because of the absence of established measures and numeric criteria for determining impairment. Both the National Research Council (NRC, 2001) and research conducted for WERF (Thornton and Freedman, 2002) have recognized the difficulty of defining impairment based on narrative criteria. The basis of this concern is straightforward. The specification of numeric water quality criteria can be a difficult process, containing numerous scientific and policy decisions. These decisions are addressed during a public participation process used to establish properly adopted numeric water quality criteria. For narrative criteria, water quality analysts are required to make the requisite decisions using best professional judgment on a case-by-case basis.

The review of over 100 TMDLs for narrative criteria conducted for this research identified a wide range of concerns in determining impairment for narrative criteria. These concerns can be divided into four primary categories:

- Over-reliance on site-specific judgment of impairment;
- Use of surrogate measures or levels that may not represent impairment of the designated use;
- Misinterpretation of data; and
- Absence of nuisance thresholds for aesthetic impairment.

Each of these areas of concern is discussed below.

3.2.1 Over-Reliance on Site-Specific Judgment

The first obstacle in determining the existence of impairment based on narrative criteria is reliance on the use of judgment in the decision-making process, caused by the lack of clearly documented translators and/or protocols. The majority (>90%) of TMDLs reviewed for this research included a *de facto* statement that the water quality was impaired, without providing data or analysis to support that conclusion. This lack of documentation may indicate that specific protocols were not followed for the determination of impairment. For those TMDLs that did state how the impairment decision was made, over half (52%) cited best professional judgment. This result is consistent with Thornton and Freedman (2002), who found that states often base listing decisions for narrative criteria on best professional judgment. Case-by-case decisions based on judgment are a concern because those making the decisions may not have the experience to exercise appropriate judgment, and site-specific judgments increase the potential for variability in how impairment is assessed at different sites.

3.2.2 Use of Surrogate Measures and/or Inappropriate Levels

The TMDL process considers a causal chain originating with the source of a stressor and concluding with impairment of the designated use (see Figure 1-2). Ideally, quantitative measures are defined at both ends of the chain, reflecting both use attainment and the degree of source control required to achieve that attainment.

The TMDL review conducted for this research (along with other studies, e.g., NRC, 2001) found that the specific mechanisms along this chain are often poorly understood for ecological impairments. Approximately 20% of the reviewed TMDLs made the determination of impairment based upon data for surrogate measures located near the "source" end of the causal chain, not direct measures of the narrative criterion. In the absence of clear source-response relationships, surrogate measures may not accurately reflect the presence or absence of impairment. Selection of appropriate levels to define impairment was a related issue, both for surrogate and direct measures of impairments. Once a measure has been selected for the determination of impairment, the next step is to define a target level for this measure that serves as the impairment threshold. The TMDL review conducted for this research found extreme variability in how target levels were selected. The biggest source of concern was selection of target levels based upon values from sites that were not representative of the waterbody of interest. For example, one state used the U.S. EPA Red Book/Gold Book (U.S. EPA, 1976; 1986) criteria for phosphorus to determine eutrophication impairment. The Red Book/Gold Book contains a single target level for phosphorus and no consideration of site-specific conditions. A better approach for determining eutrophication impairment would have been to use a nutrient target reflective of ecoregional conditions and the designated use of the waterbody.

3.2.3 Data Interpretation

Three issues were identified during the TMDL review with respect to data interpretation for determining impairment based on narrative criteria:

- Consideration of measurement variability;
- Consideration of natural factors suppressing environmental response; and
- Use of fish consumption advisories.

The determination of impairment should consider the variability and uncertainty of the test measurements, regardless of whether the criterion is numeric or narrative. The review of toxicity TMDLs showed a potential for false identification of impairment caused by incomplete consideration of measurement variability. Failure of a single whole effluent toxicity (WET) test may not indicate impairment, especially if the group of observed responses is not sufficiently greater than the natural range of response of control organisms. In accordance with WET test protocols (U.S. EPA, 1991), effluent toxicity data should be examined relative to the amount of impairment in control samples before conclusions about impairment are reached. In addition, the duration and frequency variability of the toxicity results also should be assessed when making the determination of impairment.

Another issue associated with data interpretation is the determination of whether apparent impairment is caused by a man-made stressor or by natural factors. Biological and toxicological endpoints respond to a wide range of factors, both man-made and natural. For example, a year of observed reduced fish recruitment could be caused by extreme climate events during the spawning period, or it could be caused by both natural and anthropogenic stressors, either of which might have produced the same effect alone. The measurement of an environmental response alone is not necessarily indication of impairment; the data must be properly reviewed to determine that the response is not caused by natural factors. Not only should data be reviewed for the cause of impairment as part of the listing process, but additional data collection and analysis during TMDL development may determine that the source of impairment is natural and provide support for delisting.

The final issue with data interpretation in the reviewed TMDLs was the use of consumption advisories as the sole basis for determining impairment. The threshold required to issue consumption advisories may be much lower than that that used to define nonattainment of water quality standards and impairment of the designated use. For example, shellfish consumption advisories can be issued solely because of the proximity of a wastewater treatment plant to shellfish beds or because of the absence of a sanitary survey. Such advisories, therefore, do not necessarily indicate impairment of the designated use (EPA, 2000b).

3.2.4 Nuisance Thresholds for Aesthetic Impairment

The determination of thresholds for defining aesthetic impairment was identified as a challenge in the review of narrative TMDLs in Chapter 2.0. This is because aesthetics is a human value judgment that can be dependent on an individual's perceptual "set" (an example of selective attention—Coren et al., 1984) regarding what is or is not important in a given situation. Perceptual set is defined as "a temporary readiness or disposition to perceive in a particular way" (Colman, 2001). A perceptual set may be based on an individual's newspaper reading, TV viewing, personal observations, and hearsay, for example, and predisposes the individual to respond to a new stimulus in a particular way.

The influence of perceptual set in judging issues concerning water quality has been little studied (Smith et al., 1995), but it follows that the opinions of a few individuals (experts included) cannot be assumed to be the opinion of society or even a small group of water users. For instance, perception, and therefore an individual's responses, of whether a waterbody is impaired as a consequence of lowered water clarity may well be different if the respondent fishes for trout, water skis, or simply enjoys passive recreation. The presence of litter can alter an individual's perception of an adjacent waterbody (Dinius, 1981). A water discoloration (a naturally caused yellow hue) has been shown to cause observers to "incorrectly" perceive a water body's clarity to be less than it really was (Smith et al., 1995), but only when the individuals did not know that the color was naturally caused.

The judgment of impairment and the ascribing of thresholds are complex and may need to be defined for a particular group or groups of individuals on a case-by-case basis. The determination of threshold levels may be difficult and best professional judgment is often used to list a waterbody. For example, many of the eutrophication TMDLs reviewed chose a literature value of 50 µg/l chlorophyll *a* to represent aesthetic impairment, without any consideration of site-specific conditions. Additional data collection and analysis during TMDL development may result in a revised site-specific threshold for aesthetic impairment. The waterbody could be delisted without development of a TMDL should the revised threshold be determined to not be exceeded.

3.3 Consistent Decision Process

This section provides some guiding principles to ensure a more consistent approach to the determination of impairment based on narrative criteria. They are intended to address the issues identified in the review and summarized in Section 3.2.1. The first principle is addressed to the state regulatory level, because states are ultimately responsible for determining impairment and the challenges associated with narrative criteria cannot be solved completely without statewide protocols. The remaining principles are targeted to those responsible for making site-specific determinations of impairment for narrative criteria. Case examples are used to illustrate how the principles may be applied.

3.3.1 Guiding Principles

Principle #1: States should develop protocols to allow for consistent determination of impairment based on narrative criteria.

The lack of consistency observed in the determination of impairment based on narrative criteria can be remedied through the development and application of standard protocols. Theoretically, the consistency issue could be solved through the development of properly adopted numeric criteria for all potential types of impairment. Development of these numeric criteria is not feasible in the near term. There is, however, a progression of steps that can be taken towards this end that have already been implemented by some states. Starting from the simplest, these steps are:

- Development of general guidance for the consideration of narrative criteria;
- Development of specific protocols for the assessment of each type of narrative criterion; and

- Adoption of guidance on how to develop specific numerical translators for each narrative criterion.

These steps are roughly consistent with National Pollutant Discharge Elimination System (NPDES) regulations, at 40 CFR §122.44(d)(1)(v) and (vi), which provide instructions for developing water quality-based effluent limits from narrative water quality criteria to meet the requirement from which such limits are derived and to comply with all applicable water quality standards. Section 122.44(d)(vi) requires the state to establish effluent limits for pollutants contributing to excursions above a narrative criterion based on one of three approaches: 1) a proposed state criterion or an explicit state policy or regulation interpreting its narrative water quality criterion; 2) on a case-by-case basis using U.S. EPA's water quality criteria developed pursuant to CWA §304(a); or 3) through the use of an indicator parameter for the pollutant of concern.

In addition, U.S. EPA (Grubbs, 2001) has strongly recommended that states adopt formal plans for addressing nutrient criteria. Recognizing the difficulty in immediately adopting numeric criteria, the agency recommends that states develop nutrient criteria plans to bridge the gap between U.S. EPA's statutory role of producing criteria guidance and its review and approval of standards. In these plans, U.S. EPA expects states to describe a systematic approach to assess the sensitivity of waters to over-enrichment and the need for nutrient criteria to protect designated uses. U.S EPA recommends that states describe their strategy for deriving quantitative endpoints, either as numeric water quality criteria or as detailed mechanisms for translating a narrative criterion into numeric values on a case-by-case basis.

There is a trade-off between consistency and flexibility. Any protocol developed to improve statewide consistency must still allow sufficient flexibility to consider site-specific conditions. The examples provided below demonstrate the appropriate balance between consistency and flexibility.

General Guidance

The simplest step consists of general guidance for the consideration of narrative criteria. While this falls short of criteria-specific protocols, it provides a first step towards achieving consistency in the impairment determination for narrative criteria. The Washington Department of Ecology (1997) developed a policy for determining impairment based on narrative criteria that requires that the impairment decision be based upon three pieces of information: 1) documented environmental alteration; 2) documented impairment of a characteristic use; and 3) identification of a direct human-caused contribution.

The guidance, although general, addresses many of the challenges identified previously. It requires documentation of the impairment decision, which is often not included in the final TMDL document based on the results of this study. It requires that impairment be observed on the characteristic use, and not a surrogate indicator. Finally, it requires identification of a direct human-caused contribution, avoiding the problem of identifying "impairment" in cases where natural conditions caused suppression of an environmental indicator. This guidance is discussed in more detail in Section 3.3.2 where examples of state policy addressing this issue are provided.

Criterion-Specific Protocols

The second level of statewide guidance would consist of the development of specific protocols for the assessment of each type of narrative criterion. The guidance would stop short of setting specific numeric measures, but would provide a listing of potential measures that are appropriate to the state, as well as accepted methodologies for setting target levels. An example of this type of guidance is the "Provisional Implementation Guidance for Determining Sediment Deposition Impacts to Aquatic Life in Streams and Rivers" developed by the Colorado Department of Health and Environment (2002). This guidance document is intended to bring a common approach to assessing streams for the impacts of sediment deposition across the state of Colorado.

Criterion-Specific Translators

The final level of statewide guidance would consist of the statewide adoption of translation procedures, which would provide official numeric target measures for narrative criteria. These procedures could be a mathematical loading/response model that is referenced in the state or tribal water quality standards as a "translator" of narrative criteria for water quality parameters that are not otherwise easily related to a pollution source. The use of these numeric translators is highly desirable and should be a long-term goal of all states because they would eliminate many of the challenges associated with narrative criteria. The use of translation procedures is being recommended through U.S. EPA's Consolidated Assessment and Listing Methodology (CALM) process (U.S. EPA, 2001a).

There also may be a legal need to formalize the assessment of narrative criteria with translator mechanisms. Barnes & Thornburg and Parsons (2002) cite administrative and judicial decisions that have concluded "narrative criteria should not be applied to any water in the absence of a duly promulgated numeric translator."

Principle #2: The determination of impairment should consider all relevant available protocols.

A wide range of criteria-specific protocols have already been developed, both at the state and national levels, for determining the existence of impairment based on narrative criteria. Rather than re-inventing the wheel on a case-by-case basis, those involved in making the determination of impairment should make full use of all available guidance. Specific documents of interest include:

- Draft U.S. EPA guidance on the Consolidated Assessment and Listing Methodology process (U.S. EPA, 2001a)
- Protocol for Developing Sediment TMDLs, (U.S. EPA, 1999a)
- Protocol for Developing Nutrient TMDLs (U.S. EPA, 1999b)
- Evaluation Guidelines for Ecological Indicators (U.S. EPA, 2000c)
- Rapid Bioassessment Protocols for Use in Streams and Rivers (U.S. EPA, 1989)
- Developing Eutrophication Standards for Lakes and Reservoirs. A Report Prepared for the Lake Standards Subcommittee (NALMS, 1992)
- Development and Adoption of Nutrient Criteria into Water Quality Standards (Grubbs, 2001)
- Evaluation of Sediment Transport Data for Clean Sediment TMDLs (USDA, 2000)
- Evaluating Whole Effluent Toxicity Testing as an Indicator of Instream Biological Conditions (Diamond et al., 1999)

The recommendations provided in the above guidance serve as the basis for many of the guiding principles listed below.

Principle #3: The rationale used to determine the existence of impairment should be rigorously documented.

The WERF review of the 303(d) listing process (Thornton and Freedman, 2002) found that best professional judgment was a key component to state determinations of impairment for narrative criteria. The review of narrative criteria TMDLs conducted for this research found that these judgment decisions were poorly documented, if at all. Adherence to this principle is designed to provide two benefits. First, the requirement of preparing rigorous justification will indirectly require that rigorous thought be put into the impairment decision. "Off-the-cuff" impairment determinations will be less likely if determinations need to be thoroughly documented. Second, this rigorous documentation also will provide a basis for revisiting the impairment decision in the future, should additional data be collected and/or new protocols be developed.

Documentation for impairment decisions should include:

- The specific methodology used to determine impairment;
- The data used to make the assessment;
- Specific observations or measurements that are the basis for the decision;
- A comparison of the impairment determination methodology to existing protocols; and
- Site-specific considerations, if any, that caused the impairment decision to differ from established methodologies.

Some states require this type of documentation. The State of Colorado (WQCD, 2002), for example, provides guidance as follows:

> Physical and biological assessments must be performed in accordance with scientifically sound methodologies…Assessment reports must include a statement of the observer's qualifications and should reference the protocols utilized. Any departures from referenced protocols and methodologies should be documented and the basis for any such departure addressed.

Principle #4: Consider the use of an intermediate listing category where evidence of impairment is inconclusive.

The current 303(d) decision process for determining impairment is binary; i.e., a waterbody is either listed as impaired or it is not. Recent reviews of the TMDL process (U.S. EPA, 1998b; NRC, 2001; Thornton and Freedman, 2002) have recognized that available data often are inadequate to make a conclusive determination of impairment. These reviews have all recommended that an intermediate listing category be created to identify waters where some data indicate the possibility of impairment, but a conclusive determination cannot be made. Impairments based on narrative criteria are especially prone to inconclusive determination of impairment, and are candidates for this intermediate listing category. Sites placed in this intermediate listing category should be given priority for additional near-term monitoring to confirm or refute the presence of impairment. This priority monitoring can be used to prevent further degradation in the event that impairment is confirmed, and allow scarce TMDL resources to be focused on those sites that are truly impaired.

The decision to create an additional listing category for impairments contains policy implications that are beyond the scope of this research. The guidance provided here does not explicitly call for creation of a new listing category, but does recommend that the references cited above be reviewed, and creation of a new listing category be considered.

Principle #5: If fish consumption advisories are to be used as the basis of an impairment determination, the scientific basis of the advisory should be investigated to determine that it is appropriate.

Fish consumption advisories can be issued for a wide number of reasons, and the basis of the advisory may be not be consistent with the state's water quality standards and/or listing methodologies. U.S. EPA (2000b) provides guidance on the use of fish and shellfish consumption advisories to determine impairment, and recommends that the following criteria be met prior to making an impairment determination:

- The advisory is based upon fish tissue data specific to the waterbody of concern.
- The risk assessment parameters of the advisory or classification are cumulatively equal to or less protective than those in the water quality standards.

Regional or statewide fish consumption advisories, in and of themselves, may not be sufficient to make an impairment determination. Nor are fish consumption advisories based upon risk assessments that are more protective than those provided for in state water quality standards. The presence of fish consumption advisories, while not always sufficient to serve as the basis of an impairment determination, should serve as a trigger for more detailed analysis. The quantity and quality of the data used in support of the consumption advisory, as well as any other potential related sources of impairment, must be thoroughly reviewed before the impairment determination is made.

3.3.2 Case Examples

Several states have developed guidance, policy and/or rules to provide for a consistent decision process when determining the existence of impairment based on narrative criteria.

3.3.2.1 State of Washington Policy

Washington Department of Ecology (1997) developed a policy for determining impairment based on narrative criteria as follows:

> The narrative standard for impairment of characteristic uses will be interpreted from three pieces of information:
>
> 1. documented environmental alteration using a generally accepted method based on site-specific information, with literature thresholds appropriate to the situation or with reference sites, **and**
> 2. documented impairment of a characteristic use on the same waterbody segment, **and**
> 3. identification of a direct human caused contribution to the environmental alteration.

This policy guidance, although general, addresses many of the challenges identified previously. It requires documentation of the impairment decision, which most of the TMDLs reviewed for this study did not have. It requires that impairment be observed on the characteristic use, and not a surrogate indicator. Finally, it requires identification of a direct human caused contribution, avoiding the problem of identifying "impairment" in cases where natural conditions caused suppression of an environmental indicator.

3.3.2.2 Colorado Guidance on Determining Sediment Deposition Impacts

The Colorado Department of Public Health and Environment (2002) has developed specific protocols to assess impairment from excessive sedimentation based on narrative criteria. This guidance is intended to "provide(s) an interpretation of the Colorado Water Quality Control Commission's (Commission) narrative standards as they apply to sediments which may form deposits detrimental to the attainment of aquatic life uses." The guidance does not provide specific numeric targets, but provides an assessment methodology for setting target levels. The framework consists of five steps:

1. **Identify candidate sediment-impacted segments.** This step is a screening level identification of stream reaches or segments where sediment impacts are known to occur or suspected to occur. Guidance is provided on the sources of information available to conduct this identification. For streams identified at this step, the degree of standards attainment is then determined by follow-up assessment studies according to Steps 2 through 5.
2. **Establish expected condition criteria.** This step consists of characterizing and classifying the study stream by identifying the watershed, ecoregion, flow regime, channel morphology or type, geological, physical, and other relevant chemical and biological attributes that are crucial for the selection of a matching expected condition. This information is then used to match the candidate stream to the expected condition to the maximum extent practicable.
3. **Identify expected condition.** Step 2 and Step 3 are closely related and when completed provide the expected condition that provides the basis of comparison to the specific study stream site or stream reach. In Step 3, the actual expected condition is identified through a tiered approach that can range from the use of reference sites to the use of conceptual or modeled expected conditions developed by expert consensus.
4. **Comparison of study segment with expected condition.** Step 4 provides the comparison of the stream bottom substrate habitat (as it relates to sediment deposition) and accompanying aquatic life with that of the expected condition. This process provides the information necessary to determine the percentage of expected condition for the habitat and biological metrics or assessment endpoints.
5. **Assess condition or degree of sediment impacts.** This is the final step in the process of determining the status of the aquatic life uses as impacted by sediment deposition. At this step, categories of narrative standard attainment are assigned, based on the combination of percentage of expected condition for physical habitat and percentage of expected condition for biology.

3.3.2.3 North Carolina Guidance for Eutrophication Impairment

The North Carolina Department of Environment and Natural Resources (2000) provides a good example of guidance for determining use support for eutrophication-related narrative impairment. An excerpt from this guidance is as follows:

> Several water quality variables may help to describe the level of eutrophication. These include pH, chlorophyll *a*, dissolved oxygen, phosphorus, nitrogen, turbidity, total dissolved gases, and other quantitative indicators. Some of these have specific water quality standards. But in order to appropriately evaluate the attainment of use support a clear weight of evidence approach must be used. This approach can be flexibly applied depending on the amount and

quality of available information. The approach uses multiple quantitative water quality variables, third party reports, analysis of water quality complaints, algal bloom reports, macrophyte observations, reports from lake associations, fish kill reports, taste and odor observations, aesthetic complaints, the episode frequency of noxious algal activity and reports and comments from the Wildlife Resources Commission. The weight of evidence approach must be carefully and professionally evaluated. In following this approach, use support suitability for agriculture, aquatic life propagation, maintenance of biological integrity, wildlife, recreation, water supply for drinking, culinary or food processing purposes, can be holistically evaluated.

It may be generally agreed that excessive amounts of the hypereutrophication causal variables, nitrogen and phosphorus, are the principal culprits in eutrophication related use impairment. Indeed, these causal variables are important concerns, however, climate and hydrology factors and the biological response factors (chlorophyll, phytoplankton, fish kills etc.) are also essential to evaluate because they may control the frequency of episodes for potential use impairment. The basis for regulatory control of nutrient over-enrichment must rely on biological responses to nutrient delivery as well as environmental effects. It is not appropriate to determine eutrophication related use impairment with the quantitative assessment of an individual water quality variable (i.e., chlorophyll *a*). Nor is it appropriate to utilize a fixed index composed of several water quality variables, which does not have the flexibility to adapt to numerous hydrological situations. Without presentation of detailed technical explanations and examples, it must be acknowledged that there are highly complex and dynamic ecosystem interactions which link measures of water quality variables and biological response variables to the determination of waterbody use support. And because of this dynamic complexity a weight of evidence approach must be used as a protocol in determining use support attainment utilizing all sources of readily available information.

The North Carolina guidance also provides specific case studies demonstrating the application of the guidance.

3.4 Selection of Measures and Levels of Impairment

Section 3.2.2 discussed the problem of using surrogate measures to define impairment that may correlate poorly to support of the designated use. This section provides guiding principles to ensure that appropriate measures are selected to determine impairment. These guidelines were derived from numerous sources, including U.S. EPA's Evaluation Guidelines for Ecological Indicators (U.S. EPA, 2000c), the NRC (2001) review of the TMDL process, and existing U.S. EPA (1999a, 1999b) TMDL guidance documents. Case examples are used to illustrate how the principles may be applied.

3.4.1 Guiding Principles

Principle #1: The measure used for determining impairment should be as closely linked to the designated use as possible.

This principle is designed to ensure that surrogate measures to define impairment reflect whether impairment exists. U.S. EPA (2000c) describes this concept as Relevance to Ecological

Function, stating that the relationship between the indicator and the environmental response should be clearly documented. The NRC (2001) review of the TMDL process recommends that "The measured criterion must be logically derived from the qualitative statement of the designated use."

The most direct method for assuring linkage to the designated use is to select a measure at the response end of the stressor-response causal chain. Some examples of this principle are as follows:

- For eutrophication impairments, observations on measures such as dissolved oxygen or clarity are preferred to plant biomass (e.g., chlorophyll *a*), which is preferred to observations on nutrient concentration.
- For toxicity impairment, ambient toxicity tests are preferable to whole effluent toxicity measurements.
- For ecological impairment, observations on in-stream communities generally are preferable to total suspended solids measurements.

Table 3-1 presents a range of direct measures of impairment that were used in the reviewed TMDLs.

Table 3-1. Direct Measures Used in TMDL Impairment Determinations.

Impairment Category	Measures of Impairment
Toxicological	Ambient toxicity testing; fish tissue concentration
Ecological	Fish population; fish life stage distribution; fish community structure; index of biotic integrity (IBI); EPT score; cobble embeddedness; sediment size distribution
Aesthetics	Secchi depth; turbidity; noticeable debris; presence of sludge banks
Reservoir Capacity	sedimentation rate; calculated reservoir design life
Eutrophication	Chlorophyll *a* concentration; Trophic Status Index (TSI); presence of algal blooms; %lake bottom covered by macrophytes; taste and odor problems

Indicators closer to the source end of the stressor-response chain may be used in the determination of impairment in certain situations. The conditions where these types of surrogates are appropriately used include those where:

- Sufficient data exist to define the relationship between the surrogate indicator and the designated use.
- The surrogate indicator is used in conjunction with other indicators as part of a weight of evidence approach to determining impairment.

While direct measures are preferred in making a determination of impairment, indicators closer to the source end of the stressor-response chain should be monitored throughout the listing and TMDL development process. This information will help determine likely stressors and sources, as well as set targets for TMDLs.

Principle #2: The measure used for determining impairment should be readily determined using well-accepted techniques.

This principle is designed to ensure that the methods required for sampling and analyzing the selected measure are technically feasible, appropriate, and efficient for use in a monitoring program (U.S. EPA, 2000c). The method of data collection and the required logistics for the selected measure must both be considered. Standard, well-documented methods are preferred; the use of novel measurement techniques requires explicit justification. Furthermore, the collection method should be compatible with the design of the monitoring program in which it is included. Other logistical considerations include the resources required to implement the monitoring, including time and effort required to collect, analyze, and report the data.

The NRC (2001) TMDL report succinctly states this principle as, "WQS must be measurable by reasonable attainable monitoring data."

Principle #3: The expected variability of the measurements used for determining impairment should be considered.

This principle is designed to account for the human errors of measurement and natural variability over time and space, as it is important to understand the variability in the selected measure to distinguish extraneous factors from a true environmental signal (U.S. EPA, 2000d). Quality assurance plans are required for any monitoring program, providing the means and methods to audit potential measurement error at each step. Natural variability needs to be assessed in terms of temporal (within and between seasons and years) and spatial terms. The potential for variability of toxicity also should be considered; for example, anthropogenic toxicity may be seasonal or batch-related, and so a single toxic event during a year may still be a valid observation. Finally, the discriminatory ability (i.e., the ability to discriminate differences among sites along a known condition gradient) should be considered. In summary, the selected measure should have a sufficiently high signal-to-noise ratio, such that observed changes in the measure can be directly attributed to environmental impairment as opposed to measurement variability.

As an example, certain fish species are known to undergo large natural fluctuation in population levels. Population levels of these species would be a less suitable measure for determining impairment than a species that exhibited more stable population numbers over time.

Principle #4: Measures should be consistent with existing data, wherever feasible.

If possible, the measure used to determine impairment should be consistent with existing data, as opposed to development of a unique measurement (U.S. EPA, 1999a). This poses two benefits. First, use of an existing measure will allow use of previously collected data at other sites for defining reference conditions. Second, use of an existing measure will facilitate assessment of temporal trends using historical data from the site. The qualifier "wherever feasible" is added to allow flexibility for cases where no existing measures adequately characterize the designated use. The use of new measures to define impairment is appropriate and should be encouraged, if the new measures allow for significantly better determination of attainment of designated uses. Also, consistent with Principle #2 above, new measurements and sampling and analytical methods are sometimes needed to stay current with the state-of-the-practice. For example, NTU has replaced JTU as the accepted measure of turbidity, and measurement of solids by suspended sediment concentration (SSC) may be more appropriate than total suspended solids (TSS). In some cases, old measurement techniques should be continued along with the use of new methodologies.

Principle #5: Measures should be readily understood.

A useful measure of impairment must produce results that are clearly understood by scientists, policy makers, and the public (U.S. EPA, 1999b; 2000c). These measures are to be used in a water quality management process; acceptance of impairment decisions by the various stakeholders of this process requires that they clearly understand the basis of the decisions being made. Adherence to Principle #4 generally leads to adherence with this principle. It is the development of new, unique measures that poses the greatest risk for misunderstanding.

Principle #6: Consider multiple measures using a weight-of-evidence approach.

The complexity of ecological and toxicity problems makes it difficult for a single measure to adequately reflect the presence (or absence) of impairment. The use of multiple measures provides multiple benefits towards addressing this difficulty. First, it provides for a more thorough determination of impairment, because a single measure may not account for all possible mechanisms of impairment. For example, a sole measure focused on fish spawning may not identify sediment-related impairments that occur during the rearing life stage (U.S. EPA, 1999a).

A second benefit is that the use of multiple measures can provide a more complete description of the system, "averaging" out some of the variability inherent to biological measurements.

Multiple measures are often used to confirm that toxicity test results accurately reflect biological impairment. For example, the State of Colorado's listing methodology (WQCD, 2002) states:

> Listing decisions based upon toxicity test results require that any such results be corroborated by biological information clearly demonstrating impacts to aquatic community health, composition, or productivity.

The use of multiple measures requires consideration of how to combine various sources of information into a single impairment determination. Three options exist for doing this:

- **Independent applicability:** Independent applicability presumes that an adverse impact occurs when any one of the measures predicts or indicates an adverse impact. This approach minimizes the risk of treating a truly impaired waterbody as un-impaired, but runs the risk of falsely listing a waterbody if all site-specific factors are not appropriately considered.
- **Multi-metric indices:** The multiple measures are combined into a single indicator (e.g., Karr, 1981; Karr and Chu, 1997). The use of published indices provides the potential to use existing expert judgment to combine multiple sources of information into a single decision.
- **A weight-of-evidence approach:** The weight-of-evidence approach considers all available information and bases the decision on the preponderance of the evidence. This approach provides for the most thorough consideration of all available data, but also requires the use of professional judgment by the analyst.

In considering the use of multiple measures, the TMDL developer should not delay the TMDL process if a single measurement provides obvious evidence of impairment. If some uncertainty remains, the TMDL developer should place emphasis on the need to develop measures that are interim to costly TMDLs. For example, when one or more indicators predict impairments that may not be realized by other measures, interim monitoring will assist in the proactive protection of resources. The interim monitoring allows focus to be made through additional understanding and reduced uncertainty, rather than immediately committing scarce funds toward developing a TMDL that may or may not be effective.

Principle #7: The threshold level used to define impairment should be demonstrated to be relevant to the specific designated use and local condition.

Once the impairment measure is selected, it is essential that the threshold level used to determine impairment be relevant to the site-specific use and local condition. For example, the target level of a macrobenthos index for a channelized urban stream should reflect the local habitat quality, and would differ from the target level for a pristine watershed. This concept is being investigated under the title of tiered aquatic life uses, and is discussed in Karr (2000). U.S. EPA (2002b), in their discussion of ecologically based water quality goals, states:

> Historically, water quality programs have relied upon performance-based standards to establish allowable pollutant loads, monitoring requirements, and enforcement actions. Performance-based standards focus on individual pollutants and are used to regulate specific activities. The use of impact standards or outcome indicators—requiring that a certain result be achieved—serves a complementary function that provides flexibility, realism, and a direct measure of ecological integrity. It is of critical importance that impact standards be specific, well defined, and allow for waters of different natural quality and different desired uses.

The State of Ohio has been at the forefront of the use of tiered designated uses, including use categories of Limited Resource Water, Modified Warmwater Habitat, Warmwater Habitat, and Exceptional Warmwater Habitat (Yoder and Rankin, 1995).

3.4.2 Case Example

The impairment determination for the Van Dusen River and Yager Creek, California, TMDL followed many of the guiding principles for the selection of measures described above (Van Duzen River and Yager Creek Total Maximum Daily Load for Sediment. U.S. EPA Region IX, December 16, 1999). The measures considered in the assessment were:

- **Fish population:** Population levels of anadramous fish were compared to historical levels, showing significant decreases over time.
- **Aggradation of the stream channel:** Passage of fish to spawning and rearing reaches of the stream was observed to be restricted during periods of low flow. Historical (1968) stream morphology data were compared to recent (1998) levels, and bed elevation was found to have increased on the order of two feet over that period of time.
- **Sedimentation in pools and gravels:** Impairment of spawning and rearing habitat was a suspected contributor to the decrease in fish population. Several measures were developed to assess this, consisting of % fine sediment, embeddedness, % pools, and maximum pool depth. Observed levels for all of these measures were found to be below published thresholds for healthy habitat.
- **Hillslope sediment delivery**: Several measures concerning the potential for sediment delivery to the stream were evaluated. These included forestry practices, road and skid trail density, and the condition of road crossings.

This example adheres to many of the guiding principles listed in Section 3.4. First, the measure of "fish population" is directly linked to support of the designated use. Second, the long-term trend in fish populations was evaluated, allowing full consideration of year-to-year variability. Next, the use of measures consistent with existing data allowed for historical comparisons of both fish populations and streambed elevation. Finally, multiple measures were used covering the spectrum of the stressor-response linkage. The fish population measure represents the ecological endpoint, while the middle two measures (stream aggradation, sedimentation) represented stressor mechanisms. Evaluation of hillslope sediment delivery provides for consideration of man-made causes of the stressor.

While the review of the Van Dusen River and Yager Creek TMDL indicated that many of the guiding principles were followed, the review of other TMDLs revealed that several waterbodies were listed because of nonattainment of narrative criteria for toxicity, with the measure of impairment solely being the amount of land in agricultural use in the watershed. The presumption was that the predominance of agriculture would lead to in-stream impairment by pesticides. The use of this measure to determine impairment does not follow the guiding principles listed in this chapter. The measure "amount of land in agricultural use" was not directly linked to support of the designated use in these cases. This measure is located at the stressor end of the stressor-response relationship, and no site-specific data were available to demonstrate a causal relationship.

3.5 Data Interpretation

Section 3.2.3 discussed some of the issues observed in data interpretation when making an impairment decision based on narrative criteria. The following guiding principles are designed to address those issues.

3.5.1 Guiding Principles

Principle #1: Variability in measures must be addressed.

While measurement variability is an important consideration for all impairment decisions, it is especially important for narrative criteria. The review of toxicity TMDLs conducted under this research showed incomplete consideration in TMDLs of measurement variability. Some states formally require that this variability be considered. For example, the State of Colorado's listing methodology (WQCD, 2002) states:

> In-situ bioassay test results, or other ambient toxicity test results, must demonstrate adverse effects as measured by a statistically significant response relative to a representative reference or control. Inherent variability in toxicity testing results must be adequately taken into account.

The NRC (2001) TMDL report specifically noted the importance of considering the variability in bioassessment results. Consistent with the guiding principles above, the report recommends that variability be explicitly considered in selection of measures (and sampling design) to determine impairment. To the point of data interpretation, the report recommends, "States should be required to report the statistical properties of the sample data analyses."

Finally, as Principle #6 states under Selection of Measures and Levels of Impairment, results are best interpreted as part of an approach that considers the bigger picture (i.e., what do the results predict versus what other indicators show?). A single toxicity test result significantly different from the control response may not be indicative of an actual in-stream impairment, and should be supplemented by more definitive, multiple concentration tests prior to making a decision about impairment.

Principle #2: Consider geographic and temporal relevance.

Many measures representative of biological endpoints show significant variability over space and time. This variability often will result in excursions beyond impairment thresholds, even for unimpaired waters. A single (or limited number of) point measurement(s) showing nonattainment of an impairment threshold, therefore, may not provide a definitive indication of impairment. For example, samples taken from isolated backwater areas can be expected to show conditions inconsistent with the designated use, regardless of whether impairment exists. Elevated levels of suspended solids taken during a storm event may not be representative of long-term excessive sediment loading. All data need to be evaluated to determine their geographic and temporal significance, and their overall relevance to the designated use (U.S. EPA, 1999a). Adherence to this principal often requires additional monitoring. This guiding principle is relevant for numeric criteria, but takes on heightened importance for narrative criteria because of large variability in many of the biological endpoints used in narrative criteria.

Principle #3: All potential causative factors must be considered.

Once an apparent suppression of the designated use is identified, it is important to determine whether it is a manmade impairment or some "natural" feature. As cited above, the State of Washington (Washington Department of Ecology, 1997) requires that any impairment determination include "identification of a direct human caused contribution to the environmental alteration." Additional guidance on this topic is contained in Chapter 4.0, Guiding Principles for Determining Causes of Impairment.

The importance of examining all contributions can be illustrated by several of the TMDLs reviewed that were listed as impaired because of an apparent suppression of the biological condition. More detailed review of the data indicated that natural background salinity levels caused the observed biological response, and that no true impairment existed.

3.5.2 Case Examples

Not all of the reviewed TMDLs provided good examples of data interpretation. On one estuary, for example, the determination of impairment was made because of ambient chronic toxicity measurements that showed some decreases in survival and reproduction relative to control samples. Further review of the impairment decision showed that many of the guiding principles outlined in Section 3.5.1 were not followed, leading to potentially erroneous conclusions.

First, the variability in the measure was not addressed (Principle #1). While chronic toxicity test results on ambient water for the upper estuary showed some samples with poorer reproduction and survival relative to controls, a more thorough review of the variability of the results showed no significant toxicity effects. Figure 3-1 illustrates the variability analysis conducted. This figure shows the frequency distribution of toxicity test results for ambient water relative to the control. A value of 100% indicates identical test results for the ambient water and the control. A value less than 100% indicates poorer reproduction/survival in the ambient water (relative to the control), while a value greater than 100% indicates better reproduction/survival in the ambient water. The impairment assessment focused on the presence of samples with less than 100%. Inspection of the full data set (Figure 3-1) indicates a symmetrical distribution around the

100% measure. When all data were considered, the toxicity response in the ambient water was not significantly greater than in the control samples.

Second, toxicity tests on samples from the middle section of the estuary showed significantly poorer reproduction of *Ceriodaphnia dubia* in the ambient water than in the control samples, contributing to the impairment determination. Further review of these data indicated that all potential causative factors were not considered (Principle #3). The data indicate that essentially all of the toxic responses were observed during periods of elevated salinity in the middle estuary. All control tests were conducted using fresh water, and did not exhibit this effect. The role of salinity in explaining the observed toxicity results was confirmed by plotting toxicity test results against the conductivity of the sample (which is directly related to salinity), demonstrating the strong positive correlation. In this situation, the apparent "impairment" likely was caused by conducting freshwater toxicity tests using ambient water with high natural salinity, and was not a true impairment.

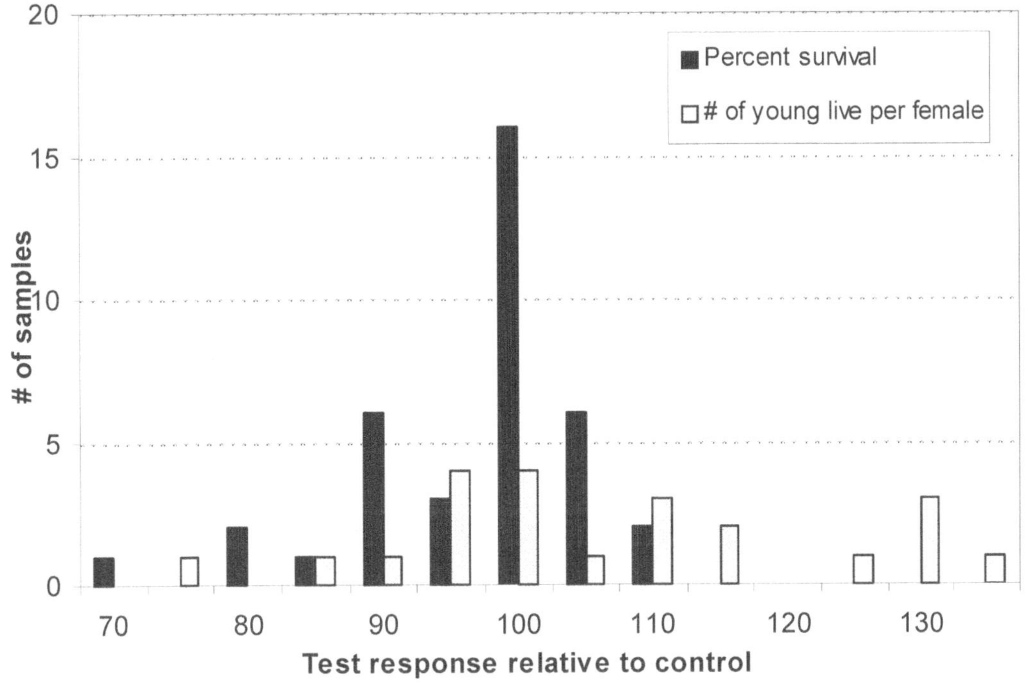

Figure 3-1. Assessment of Toxicity Test Variability.

3.6 Determining Aesthetic Impairment

Section 3.2.4 discussed the difficulties encountered in the determination of thresholds for defining aesthetic impairments. This is because aesthetics is a human value judgment and is somewhat dependent on an individual's perception. The following guiding principles are intended to address this challenge in the TMDL process for narrative aesthetic criteria.

Beyond the guiding principles provided below, it must be recognized that the assessment of human perceptual requirements is a relatively new field and with little to go on in the literature. A recent New Zealand study (Smith et al., 1995) may provide some guidance for visual clarity and hue requirements for bathing and aesthetic water uses, but the study findings have not been validated for the United States. The advice, at an early stage, of a social scientist, psychologist, or a water quality scientist familiar with the issues in the approach and survey questionnaire design, is recommended.

Principle #1: Assess the potential causes for aesthetic impairment.

Many features of a waterscape may be important in the perception of a waterbody and hence the potential causes of aesthetic impairment of that water including, for instance, the presence of bankside litter or the condition of the riparian zone. People tend to perceive what they expect to see (Smith et al., 1995), and the surrounding area may play an important part in the judgment of a waterbody's suitability-for-use. It has been shown that the environments that recreational planners strive to define objectively are not necessarily those "seen" by the general public (Knopf, 1983) so there could be some surprises.

An assessment of the potential causes of impairment could be carried out initially by best professional judgment, but a waterside survey is preferable to discover users' concerns. After all, it is the *users* who determine the nature of the aesthetic impairment. The terms used by the public may differ from those used by scientists, and the same words and phrases may have different meanings when used by the public and scientists.

Principle #2: Develop a measurement procedure for each impairment cause.

For some attributes, measures of aesthetic impairment may be straightforward; for example, the number and size of pieces of litter per unit area. For visual water clarity and hue, appropriate measurement methods have been developed in the context of water use impairment (Smith et al., 1995). Subjective scales may need to be developed for attributes such as scum and odor.

Principle #3: Consider local input to define impairment.

Local input can be obtained using appropriately designed panel tests with user groups, supplemented with photographs (e.g., showing varying amounts of litter) or, preferably, by in-situ measurements of the potential impairment causes and perceptual assessments at the sites of concern. It may be necessary to conduct studies at a variety of sites covering a range of each impairment type (e.g., visual clarity) so that a range of perceptions can be obtained. This is necessary to define thresholds, not just whether a particular waterbody is impaired. The general population or defined user groups can be targeted for questionnaire assessment, as appropriate. Surveys of local communities have been used successfully to determine impairment thresholds for lake eutrophication (Heiskary, 1989; Heiskary and Wilson, 1989; Smeltzer and Heiskary, 1990).

For measurements such as visual clarity, it is essential to make the measurements as close to the interview location as possible; it is of little value, for instance, to measure a lake's clarity in the middle of the lake if perception assessments are made at a beach some distance away. It seems likely that one of the main attributes of aesthetic impairment will be reduced visual clarity, and the potential effects of perceptual set will need to be borne in mind during questionnaire design and field study.

CHAPTER 4.0

GUIDING PRINCIPLES FOR DETERMINING CAUSES OF IMPAIRMENT

4.1 Introduction

Once a state has determined that a waterbody is failing to meet its designated uses, a coherent process is needed to determine the causes of that impairment. Determining causes often is complicated by the presence of a number of candidate causes, two or more of which may be acting in concert to produce the impairment. If causes of impairment are not identified correctly, TMDL controls may be aimed at the wrong issues or at an incomplete group of causes. When this occurs, the implementation of TMDL controls will not result in attainment of the narrative criteria supporting the designated uses, potentially wasting time and money, and also producing disillusionment among the communities affected.

This chapter contains guiding principles to improve the determination of the causes of impairment, utilizing a step-wise sequence consistent with U.S. EPA (2000d). The relevant steps for determining the causes of impairment include:

- List candidate causes
- Analyze the evidence
- Identify potential causes (eliminate, diagnose, compare strength of evidence)

This sequence recognizes that few issues in science are certain, particularly when dealing with processes and effects in the real world rather than controlled laboratory experiments. For this reason, care is needed to ensure that sound conclusions are reached. The first step in accomplishing this is including all possible candidate causes. The next step includes progressively eliminating implausible items and re-checking the findings. Inevitably, this means that meeting a strict set of requirements may not be achievable in environmental studies, and so a "softer," wider set of criteria apply. In developing these ideas, particular attention is paid to the requirements of establishing cause-and-effect in the last step of this process, identifying potential causes. The use of a series of logical criteria for determining likely causality mirrors the approach used in human epidemiology, and in fact has been termed "eco-epidemiology" (Bro-Rasmussen and Lokke, cited in Fox, 1991). A recent set of these causality criteria (U.S. EPA, 2000d) are presented as a template for discussion.

This section contains guiding principles intended to improve the determination of the causes of impairment. It begins with a discussion of the challenges and concerns associated with determining the causes of impairment. These challenges and concerns were identified in the review of narrative TMDLs presented in Chapter 2.0. Guiding principles are then presented, categorized by the challenges and concerns they are intended to remedy. This chapter is divided into sections:

Challenges and Concerns (Section 4.2)
Listing candidate Causes (Section 4.3)
Analyzing the Evidence (Section 4.4)
Identifying the Causes (Section 4.5)

4.2 Challenges and Concerns

The review findings presented in Chapter 2.0 showed that many narrative TMDLs do not employ, or at least do not document, a sequential process to identify probable causes of impairment. Yet were they to do so, a number of benefits would result:

- Better targeted load allocations and management controls.
- More community support for the findings (by demonstrating that some activities that could initially appear to be causing the impairment may be ruled out).
- Setting precedents for other TMDL efforts.

Specific challenges that were identified in the review of narrative TMDLs associated with determining the causes of impairment include:

- Complex and poorly understood cause-and-effect mechanisms;
- Determining cause based on ambient toxicity data; and
- Quantifying relationships between loading and ambient concentrations.

Determining the causes of impairment presents significant challenges for narrative TMDLs because of complex and typically poorly understood cause-and-effect relationships between stressors and ecological responses. The review findings showed that the cause-and-effect linkage was not considered for those toxicity TMDLs where impairment was caused by a single discharge of whole effluent toxicity to an effluent dominated stream, because no ambient impairment was ever measured. Also, the review revealed that no attempt was made at these sites to identify the specific contaminants responsible for the toxicity. The one TMDL that considered multiple sources of toxicity was unable to resolve the individual contributions of toxicity from the contributing sources. Also, certain chemicals (such as mercury) undergo complex speciation reactions that affect bioavailability. For chemicals such as mercury, the source can play a direct role in its ultimate bioavailability, and therefore complicate the identification of the true causes of impairment.

For the majority of fish and benthos impairments reviewed, excess sediment load was identified as the stressor of concern. Only one-third of the TMDLs reviewed conducted an explicit assessment of the cause of impairment, either in terms of documenting that sediment was the primary stressor or defining the primary source of the excess sediment. The remaining two-thirds of these TMDLs listed the cause of impairment as known, without providing supporting data. The remaining non-nutrient ecological impairment TMDLs also lacked explicit documentation of the presumed cause-effect relationship (Section 2.4.2).

4.3 Listing Candidate Causes

The first step in determining the causes of impairment is to generate a list of candidate causes. This section provides guiding principles for listing candidate causes of impairment. They are intended to address the challenges identified in the review of narrative TMDLs and summarized in Section 4.2.

4.3.1 Guiding Principles

Principle #1: Develop a wide initial list of candidate sources and causes—include all reasonably possible causes.

Listing potential causes of impairment should be a comprehensive and exhaustive exercise. Available evidence should be collected from the current and similar situations, including all existing lists of candidate sources and causes. No candidate cause should be eliminated at this stage. The list should take full cognizance of the effects identified when the impairment was defined. Examples of effects include loss of species, shifts in abundance, and changes in physical/chemical condition.

Existing lists (e.g., from results of monitoring programs of state agencies, utilities and volunteer monitoring programs) and knowledge of upstream pollution sources and activities will be helpful in developing the list. Available public information on upstream wastewater discharge permits, limits, and violations of limits also should be reviewed and considered. Simple sanitary surveys may be helpful in identifying unsuspected sources (e.g., drains) that often are overlooked in pollution investigations.

Principle #2: Develop an initial conceptual model indicating possible links between causes and impairment.

More than one stressor may need to be present before an effect can occur; e.g., elevated nutrients *and* removal of bank-side vegetation causing undesirable aquatic growths. If a streambed is altered both chemically and biologically, full restoration may require that the cause of sedimentation as well as the store and supply of contamination be controlled. Therefore, listing of candidate causes of impairment should be accompanied by "thought experiments" using conceptual models to identify possible pathways from causes (stressors) to effect (impairment). Conceptual models also may indicate additional types of data that may be helpful.

Suggesting a possible cause is not sufficient. A conceptual model is needed that lists the cause and effects along with other candidate causes and their potential effects. Applying ecological risk assessment processes to collect, organize, analyze, and present scientific information can improve the decision-making process for the structure of the model. Figure 4-1 illustrates the process required to assemble and consider relevant information from technical and local sources. Development of a conceptual model then requires consideration of the nature and sources of potential stressors and their resulting effects on the environment (e.g., Figure 4-2 illustrating stressor effects on freshwater mussel and fish species).

The case example presented in Section 4.3.2 gives a more detailed example of a conceptual model for forestry impacts on stream benthic invertebrate communities. Although the initial model may be relatively simple (e.g., Fig. 4-3), it inevitably becomes more complex after extensive discussion with other professionals and resource users. The complexity of such conceptual models may limit their usefulness for some readers. However, some level of complexity is necessary to begin to understand a functioning ecosystem. The model may subsequently be simplified as more data and understanding of the aquatic system and its responses to pressures become available, but it is a mistake to simplify the model too early in the process of identifying causes. Furthermore, new techniques, particularly Bayesian Belief Networks (BNN), enable a reasonably complex mix of causes and effects to be modeled with less effort and more flexibility than is the case with deterministic models (Section 4.3.2).

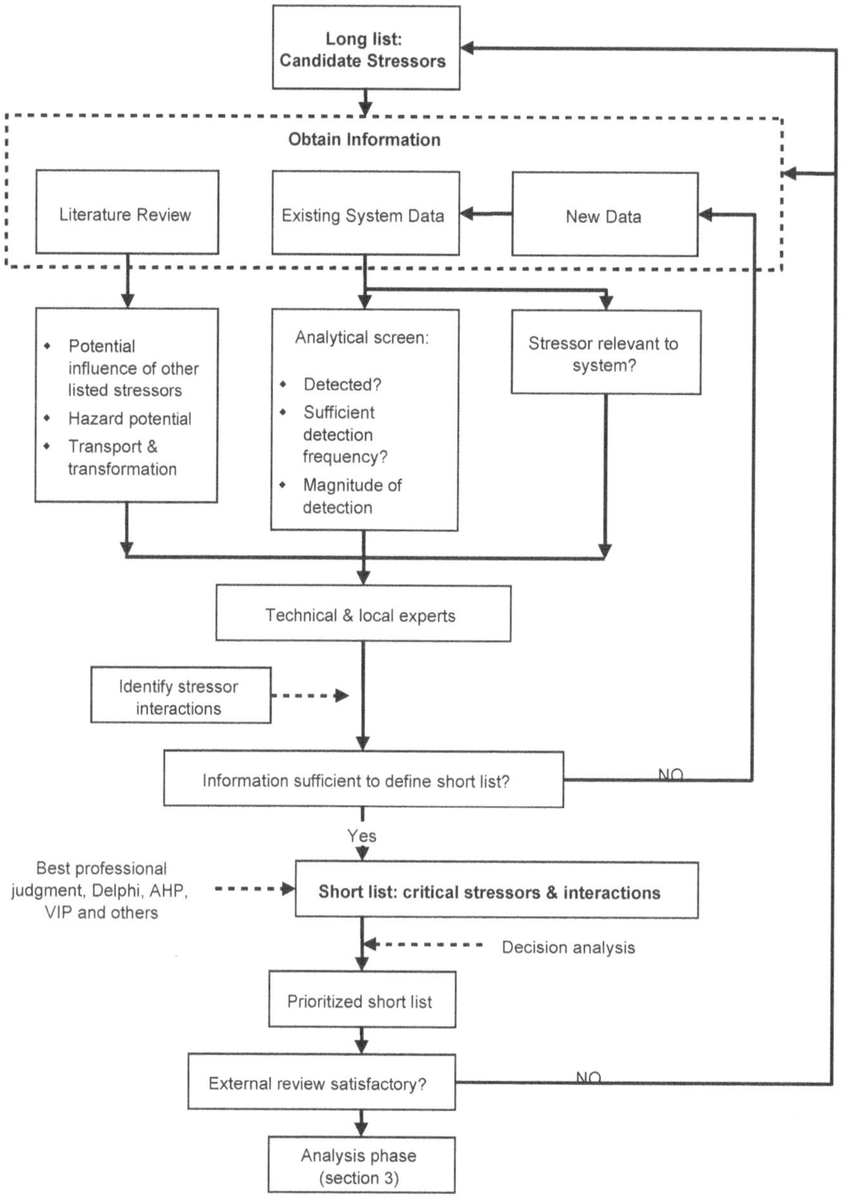

Figure 4-1. Integrated Procedure for Considering Multiple Stressor Effects on Ecosystems.
(Reprinted with permission from Foran and Ferenc, 1999. Copyright 1999 Society of Environmental Toxicology and Chemistry).

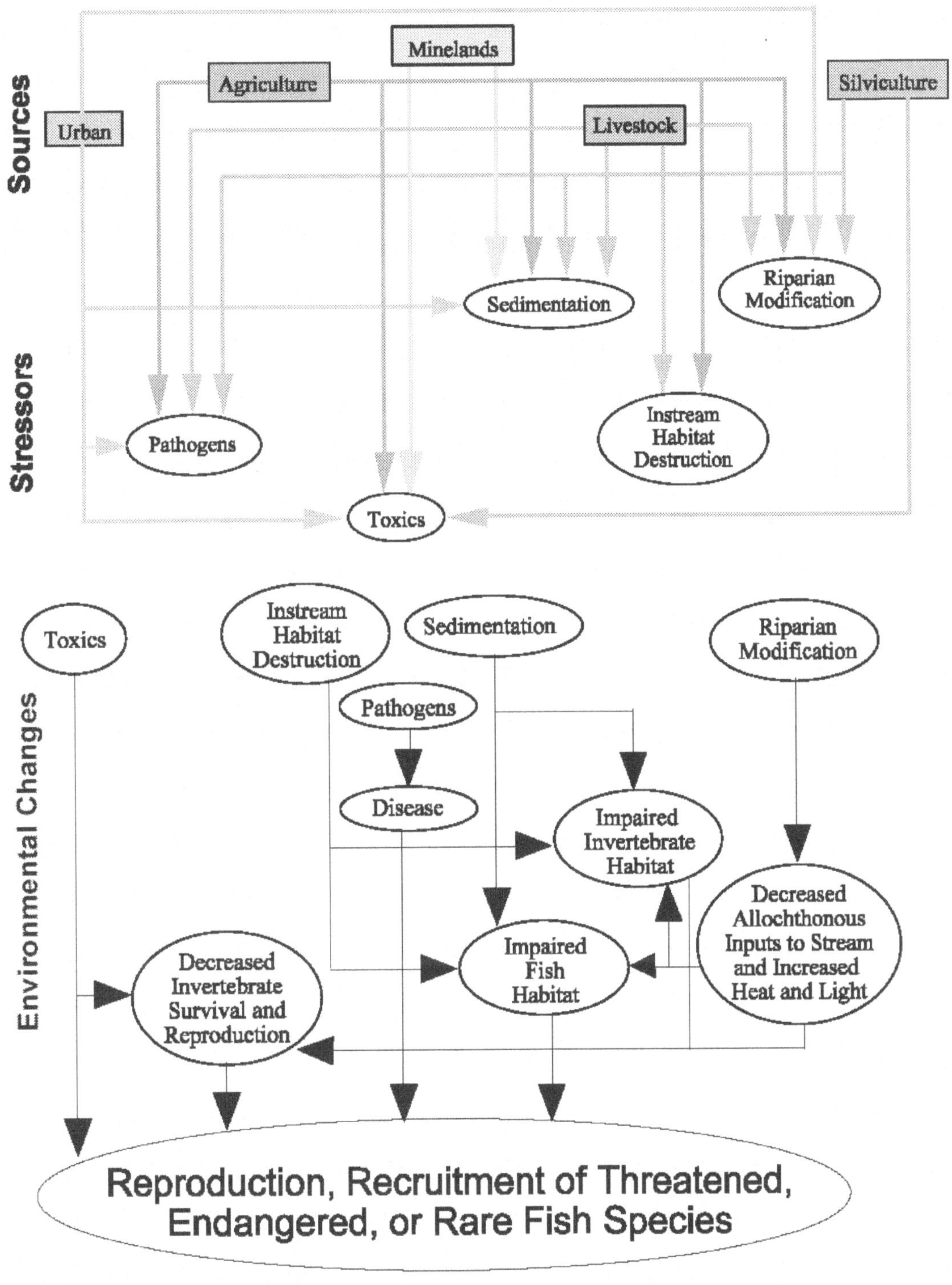

Figure 4-2. Conceptual Models for Risks to Freshwater Mussels and Fish Developed for the Clinch and Powell Valley Watershed in Virginia.
(Reprinted from U.S. EPA, 2002c).

4.3.2 Case Example

The example given herein examines how forestry management practices can affect stream health, as assessed by the presence or absence of mayflies, snails and caddisflies.[1] An initial conceptual model is shown in Figure 4-3. Following a series of consultations between ecologists, geomorphologists, foresters, and the modeler (an ecologist with experience in the ecology and hydrology of forestry-impacted streams), an expanded conceptual model was developed (Figure 4-4). The more detailed model includes interactions between such activities as riparian harvesting, roading, landings and bank erosion.

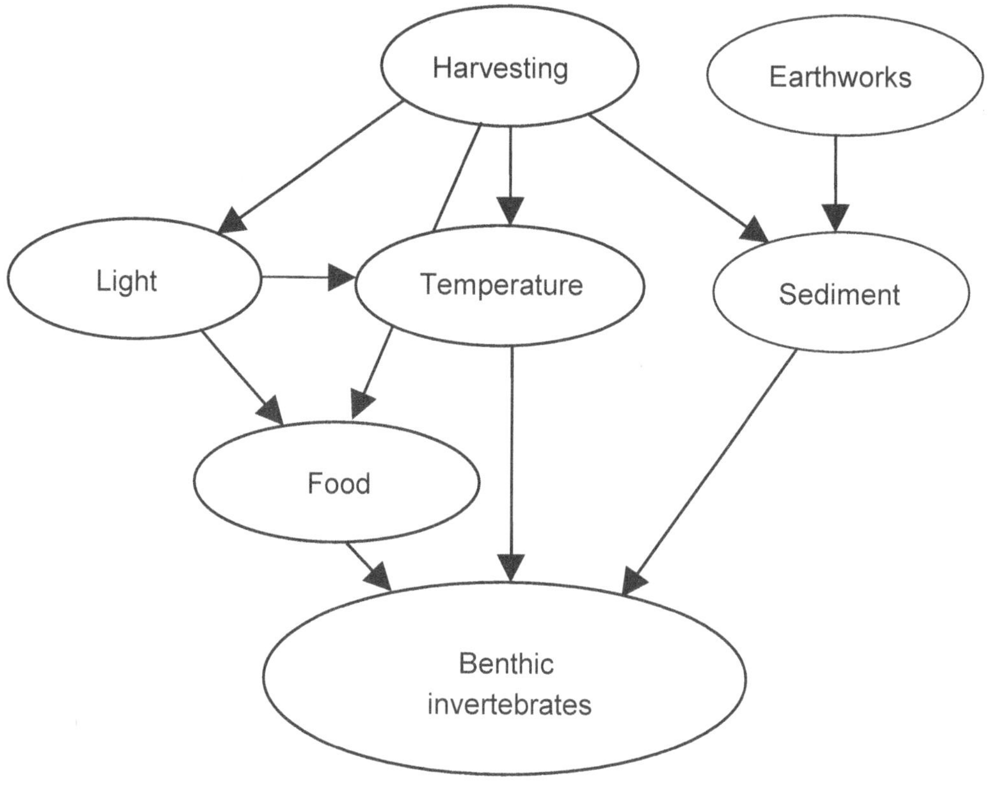

Figure 4-3. Conceptual Outline of BBN Model for Forestry Effects on Stream Health.

[1] For details see http://www.niwa.co.nz/pubs/wa/09-1/newsforum.htm#item5

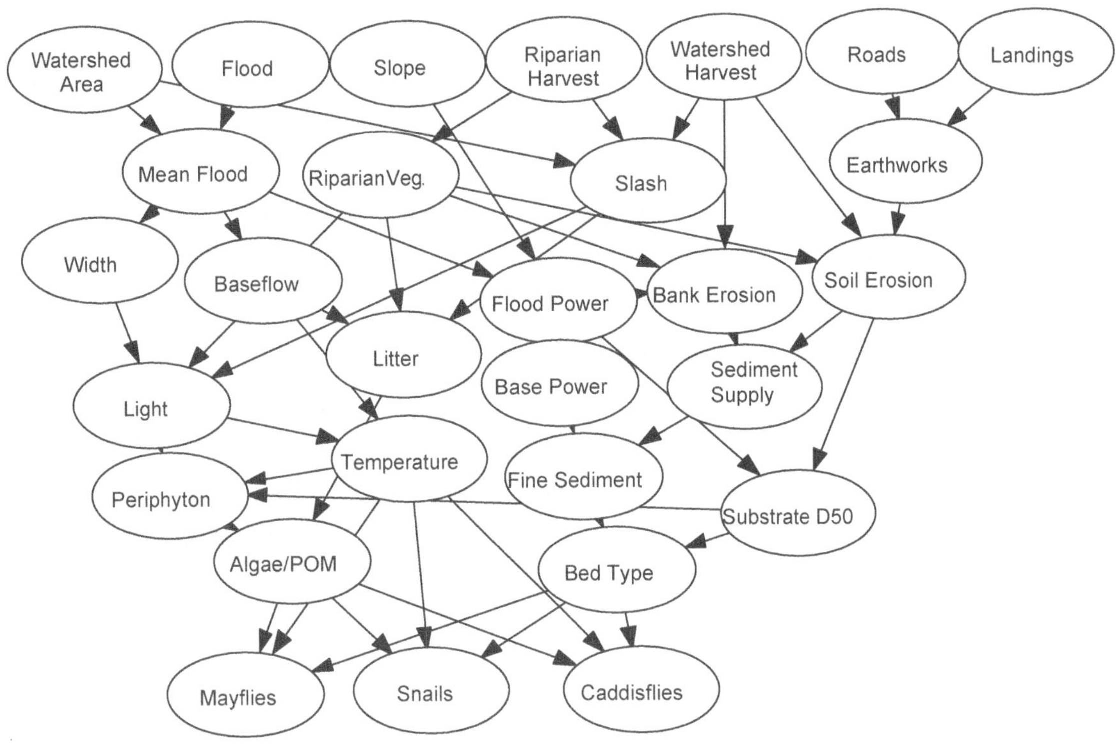

Figure 4-4. BBN Model for Forestry Effects on Stream Health. (All Arrowed Links Have Quantitative Conditional Probabilities.)

Note that this approach could be used to assess the main drivers for sedimentation or temperature, in which case the ecological endpoints need not be included in the scheme. Despite its complexity, such a model serves two important roles: 1) it represents the major links in a complex system, and 2) its preparation—requiring consultation with the "major players" in the field—serves as a means of elucidating those links. If these links are not considered and made explicit there could be a danger, for example, of inferring that ecological communities have been impacted by sedimentation resulting from forest harvesting, whereas associated roading may have been the dominant source of sediment. New, relatively simple procedures are now available to use such conceptual models in quantitative probabilistic ways—using Bayesian Belief Networks.[2]

If an attempt is made to codify the understanding of processes operating in each link, even if an "off-the-shelf" model is available, considerable complexity and effort is required. However, BBNs replace that complexity with assignment of conditional probabilities for each link. That is, a BBN is an expert system in which the cause-effect relationships between variables are defined and conditional probabilities are set for each variable in each of its possible

[2] These are Bayesian in the sense that probabilities assigned to an item may be modified at any time by new data, using Bayes' rule. While the rule itself is uncontroversial among statisticians, the assigning of "probability" to "belief" can be very much so. In the context of BBNs there is in fact little controversy, as the network, by its very title, makes it plain that one is dealing with beliefs. However for some procedures, such as formal comparisons, the controversy is more substantial with some statisticians preferring a "half-way house" using likelihood (Royall, 1997) or Bayes factors (Goodman, 1999) which essentially avoid any *absolute* measure of support for hypotheses, relying instead only on *relative* measures.

states (generally three or four per variable), given the state of "parent" variables. When evidence is presented to the network on state of one or more variables, the probabilities (beliefs) in the states of all the other variables are propagated and updated. That is, the "plausible reasoning" inherent in such networks provides a coherent vehicle for accumulating scientific knowledge (expressed as a belief probability), which many argue is a cornerstone of successful science enterprise (Nelder, 1999). Both diagnostic and prognostic reasoning are included in this scheme. This can include the case where certainty is achieved in a variable (e.g., if after first developing such a model, water temperature records are obtained). Once that happens the beliefs associated with remaining states become better defined. Finally, some quantification of belief in this scenario means that uncertainty is recognized (as recommended for TMDL models by NRC, 2001).

The model shown in Figure 4-4 was developed for data from 46 New Zealand streams (Coromandel Peninsula) during eight years of logging operations (Quinn and Walley, 2002). Implementation software is readily available, some of which (e.g., WinBUGS[3]) is freeware. The model has been used to assess the merits of controlling riparian harvesting on stream health. It and others relating to the influence of riparian buffers on the effects of logging on fish are being used in development of a decision-support system for New Zealand plantation forests.

4.4 Analyzing the Evidence

Once the candidate causes are identified, the next step in defining causes is to review the available data to provide a basis for eliminating some of the potential causes. Two guiding principles are provided in that regard.

Principle #1: Analyze available data and literature to refine the conceptual model.

Information for each of the candidate causes is critically reviewed, using available data and literature information. This does not yet constitute evidence of causation. It is merely an elaboration and documentation of the conceptual model contemplated when making the list. Four tests or steps are appropriate. These are:

Field Observations

- Are measurements of candidate causes (or surrogates) associated with observed effects by co-location, co-occurrence, mutual exclusion, or a correlation in intensity? These observations need special care in the case of mobile organisms.
- Are the effects observed in response to the mitigation or manipulation of causes? An example might be improved Biotic Index and fish recolonization observed after decreased organic enrichment.

Laboratory/Literature

- Are measurements of site concentration similar to levels where effects were observed in laboratory studies (e.g., ecotoxicology)?
- Are measurements observed that indicate specific causal mechanisms (e.g., a biomarker as an intermediate step in toxicity; a shift in species distribution associated with a new predator)?

The result of this process should be an enhanced conceptual model of the relevant causes and effects and an ability to assign some degree of confidence to its interacting components.

---

[3] http://www.mrc-bsu.cam.ac.uk/bugs/winbugs/contents.shtml

Principle #2: Assign uncertainties to the conceptual model components and predictions, and use the model to guide future data collection programs.

The conceptual modeling effort also should help to guide future data collection programs by indicating what data are required to reduce uncertainty in the model. Specifically, the uncertainty in each link of the conceptual model should be estimated, based on the availability of site-specific data. The processes that have the greatest uncertainty are primary candidates for additional data collection. The overall magnitude of the process should be considered in this decision, along with the uncertainty. For example, a process may have an uncertainty of +/-100%, but if the upper bound estimate of its magnitude still has a minor contribution to the overall impairment, the process may be dropped from consideration without additional data collection.

4.5 Identifying Potential Causes

After analyzing the evidence, three steps remain in determining the causes of impairment: eliminate, diagnose, and compare the strength of the evidence. These are in order from the most conclusive (eliminate), to the least conclusive (strength-of-evidence). The last of these includes features of the first two, as will be seen.

4.5.1 Guiding Principles

Principle #1: Eliminate candidate causes that fail logical tests, but do not rely solely on statistical null hypothesis tests.

This step relies on *logical* elimination by deductive processes. The best example is "temporality." If an effect precedes a candidate cause in space (e.g., it is upstream) or in time (occurring earlier than the cause onset), it may be logically eliminated.

Logical elimination appeals to the Popper/Platt view of science, in which hypotheses can only be rejected (Platt, 1964; Popper, 1968); if hypotheses fail to be rejected when repeatedly tested, they gain credibility. But in so doing, disproof should not rely on statistical hypothesis testing alone (especially if two-sided tests of "point-null" hypotheses, such as *t*-tests and ANOVAs are employed). This is an important and seldom understood issue that is receiving increasing attention in statistical and medical literatures (Goodman, 1993; Royall, 1997; Harlow et al., 1997; Goodman, 1999; Nelder, 1999) and in environmental work (Germano, 1999; McBride, 2002).[4] In summary, rejection of such hypotheses does not necessarily demonstrate that a strong association exists, nor does failure to reject them constitute logical grounds for accepting them as being true.[5]

---

[4] If a test of the usual "point-null" hypothesis does not result in that hypothesis being rejected, one does *not* then have grounds to accept it; it may merely mean that too few samples were taken (Royall 1997). Such hypotheses posit that there is no association *whatsoever* between a stressor and an effect; this is generally untenable, though it is assumed to be true when calculating the test's *p*-value! (*p* is the probability of obtaining data at least as extreme as has been obtained *if* the tested hypothesis is true; if it is "small"—less than the significance level—the hypothesis is rejected; otherwise the hypothesis is not rejected).

[5] The power to reject a point-null hypothesis generally increases as the number of samples increases, to the extent that a test with a large number of samples can routinely detect a trivial correlation while the same test with a small number of samples can fail to detect an important one. This is because the hypothesis being tested is in fact barren—it is false. In contrast, one sided tests of inequalities or two-sided tests of intervals do not share this feature—increased sampling can in fact strengthen the evidence for or against the tested hypothesis (McBride, 1999). For further discussion, see Griffith et al. (2001).

If no strong grounds for elimination arise, the potential cause should be retained.

Principle #2: ***If more than one candidate cause remains after the elimination step, subject them to a diagnostic analysis by seeking evidence or lines of argument that confirm a candidate cause (particularly for pathogenic organisms, chemicals, and physical properties showing repeatable behavior such as diurnal dissolved oxygen and pH).***

If more than one cause remains after elimination considerations, diagnostic analysis should be attempted. In contrast to elimination, this step relies on positive evidence (e.g., as in medicine, when a particular symptom is present). Thus, it tends to apply only to cases where there is good background knowledge and high specificity and protocols and organisms for investigation (e.g., fish kills), some pathogens (e.g., tests for animal versus human origin), contaminant chemicals, or measurements of some well-understood intermediate levels (e.g., minimum dissolved oxygen conditions occurring at multiple sites just before or after dawn strongly suggesting the importance of aquatic plant metabolism). Diagnosis is more problematic for most other stressors and for effects on non-vertebrate organisms (U.S. EPA, 2000d).

An additional example of this analysis concerns fish kills, where examination of fish tissue can confirm the presence of a particular toxicant pollutant that may have been discharged to the waterbody. Another is the detection of quantities of *Cryptosporidium* oocysts in iced drinking water, where that water was taken from a watershed during a flood event washing off farm runoff with the result being a disease outbreak in the community served (e.g., the major Milwaukee outbreak in 1993; MacKenzie et al., 1994).

The protocols associated with Toxicity Identification and Evaluation (TIE) program may be applied to better identify candidate causes of toxicity, but not all effects of concern (e.g., tumors) occur in these tests. Additionally, there may be questions concerning the sensitivity of the laboratory tests relative to field effect levels, and the appropriateness of the laboratory test species and test durations relative to natural field exposures. For these reasons, TIE procedures are considered by EPA to be part of the "strength-of-evidence" analysis (U.S. EPA, 2000d).

Principle #3: ***Always perform a strength-of-evidence analysis, even if previous steps have identified causes (in which case the analysis serves as a check on that identification).***

If elimination and diagnosis do not identify the cause(s), a strength-of-evidence analysis is required. The aim of this analysis is to find the cause(s) best supported by the evidence. Even if the eliminate/diagnose steps have apparently revealed the cause(s), it is desirable to complete a strength-of-evidence analysis as a check on those outcomes. Note too that some items in such a strength-of-evidence analysis will already have been considered in previous steps (i.e., "Temporality" in the Elimination step, and "Consistency of association" and "Specificity" in the Diagnosis step).

Principle #4: ***Complete a strength-of-evidence analysis by considering the following 11 Bradford-Hill criteria: Co-occurrence; Consistency of association; Coherence of association; Specificity of association; Temporality; Biological gradient; Biological plausibility; Experimental evidence; Predictive performance; Analogy; and Exposure indicators.***

For many years the association of effects with causes has been a central question in scientific and logical thought. In particular, a logician would see the unequivocal demonstration of cause-and-effect in the following way:

- To be considered a cause, a stressor must be a necessary and sufficient condition for the effect to occur.
- This implies that there must be a one-to-one relationship between the stressor and its effect.

A major meta-analysis of a number of studies on the possible effects of smoking on human health resulted in the development of criteria for demonstrating cause-and-effect (U.S. Dept. of Health, Education & Welfare 1964, cited by Fox 1991) (Table 4-1).

Table 4-1. Example of Meta-analysis Criteria for Establishing Cause-and-Effect Relationships.

Evidence Criterion	*Rationale (our comment)*
1. Consistency of association	Does the association occur at other times and places?
2. Strength of association	Is the disease in the exposed and not in the unexposed?
3. Specificity of association	Weak if multiple diseases can be caused by exposure to one pollutant
4. Temporal relationship	Does the cause precede the effect?
5. Coherence	Corresponds to current understanding?

Further development of these criteria concerning health effects in the workplace was conducted by Sir Austin Bradford Hill of the British Royal Society of Medicine (Hill, 1965) (Table 4-2). More recently, additional criteria have been proposed, including "predictive performance" and "probability" (Susser, 1986) and "exposure indicators" (e.g., body burden of contaminants; physiological response by production of detoxification enzymes). U.S. EPA (2000d) proposes 12 evidence criteria (Table 4-3) that incorporate many of the previously proposed criteria. Guiding Principle #4 recommends that a modified set of Bradford Hill criteria (Table 4-4) be used to evaluate the strength of the evidence on causal relationships.

Table 4-2. Example of Bradford Hill Criteria Table.

Evidence Criterion	*Rationale (our comment)*
1. Strength (of association)	Based on relative risks, ignoring sampling variability. (But small risk might arise from small gradient.)
2. Consistency	Repeatedly observed?
3. Specificity	Weak if multiple diseases can be caused by exposure to one pollutant
4. Temporality	Cart before the horse?
5. Biological gradient	One should invariably seek it
6. Plausibility	Depends on knowledge of the day (positive).
7. Coherence	Should not seriously conflict with known facts (negative)
8. Experiment	Can one manipulate the environment?
9. Analogy	Do similar causes give rise to similar effects? (accepting slighter evidence)

Table 4-3. Example of EPA-Modified Bradford Hill.

Evidence Criterion	*Rationale (our comment)*
Strongest	
1. Co-occurrence	Co-location (effect downstream of cause)
2. Temporality	Cause precedes effect in time *(necessary* for causality)
3. Biological gradient	Effects increase with exposure
4. Complete exposure pathway	Getting from A to B
Data or similar situations	
5. Consistency of association	Repeated observations
6. Experiment	In lab or controlled field trials
Combining other knowledge	
7. Plausibility	Makes sense?
8. Analogy	Similar to other well-established cases?
9. Specificity of cause	Specific relationship (tumors and PAH)
10. Predictive performance	Do effects predicted given presence of stressor actually occur? (as suggested by Susser 1986)
Relationships with all available lines of evidence	
11. Consistency of evidence	Consistent with all available evidence? (rather similar to Plausibility)
12. Coherence of evidence	Can conceptual or mathematical model explain apparent inconsistencies (e.g., fish colonize a contaminated site but breed elsewhere)?

Table 4-4. Recommended Criteria and Weightings for Strength-of-Evidence Analysis. (Modified from Fox 1991 and U.S. EPA 2000d)

Criterion	Possible Result	Score[1,2]
1. Co-occurrence[3]	Compatible	1
	Uncertain	0
	Incompatible	-3
2. Consistency of association	Always	2
	Often	1
	Seldom (or indiscernible)	-1
3. Coherence of association	Data agree with current understanding	2
	Data equivocal	0
	Data highly conflicting	-3
4. Specificity of association	Only one possible	2
	One of a few	0
	One of many	-3
5. Temporality	Compatible, uncertain, incompatible	2
	Uncertain	0
	Incompatible	-3
6. Biological gradient	Strong	2
	Weak	0
	Present but in wrong direction	-2
7. Biological plausibility	Strong	2
	Unknown	0
	Conflicting	-3
8. Experimental evidence	Concordant	3
	Ambiguous	0
	Discordant	-2
9. Predictive performance	Confirmed	2
	Ambiguous	0
	Failed	-1
10. Analogy	Many positive analogies	2
	Few or unclear	0
	Many negative analogies	-2
11. Exposure indicators	Strong evidence	3
	Weak evidence	1
	None found	-1

1 There may be No Evidence (NE) or the consideration may be Not Applicable (NA).

2 One could adopt an intermediate score; e.g., if a mild biological gradient was present a score of "1" could be assigned.

3 Positive co-occurrence scores only a 1 because it is a necessary but not sufficient cause. (In reality, the –3 score would not be applied, as this assignment would have caused the candidate to be eliminated earlier.)

Principle #5: Use a weighting matrix for candidate causes to compare the relative strengths of evidence for them.

Having completed an analysis of cause-effect data using the criteria in Table 4-4, the various findings must be weighed to arrive at overall conclusions. The recommended weighting matrix in Table 4-4 is based on material presented by Fox (1991) and U.S. EPA (2000d).

Once these scores are assigned, they can be used to make *relative*, not absolute, judgments about probable causes of impairment. In addition, scores taken from the same table are only comparable between candidate causes when there is the same type of evidence available for all of them. Weight-of-evidence analysis often is complex and not easily captured by a simple index. Nevertheless, performing comparison of the weightings will be an instructive exercise in the overall weighing of evidence. That final weighing is analogous to that performed by a judge at a trial, when presented with an array of evidence of varying degree and quality, and must be done by environmental professionals. Even so, the judgment process can include opportunity for engagement with the affected community in the watershed, by demonstrating an open approach to possible causes and then proceeding through an agreed set of criteria before coming to conclusions.

4.5.2 Case Examples

The TMDL documents reviewed cover a range of depths of analysis for determining causes of impairment. In general, they do not present an analysis following a sequence such as the guiding principles for determining causes of impairment presented above; none do so completely. As such, it is not particularly instructive to examine those that fall rather short of the considerations detailed herein. Instead, examples are presented that illustrate application of some of the principles.

4.5.2.1 Noyo River, California

The Noyo River was listed on the 1998 CWA 303(d) list by the State of California. The listing process had determined that sedimentation throughout the watershed was the cause of the decline in the watershed's cold water fishery (including migration, spawning, and reproduction of salmonids). Although the Noyo River TMDL (U.S. EPA, 1999c) noted that salmonids have been declining in recent years, the TMDL did not document the determination of causes of the impairment. The TMDL document does not evaluate the possibility that the decline in the fishery could have been caused by chemical contamination, nor was a conceptual model developed.

With respect to sedimentation, a detailed examination of candidate sources of sediment is presented in source and linkage analyses. The analysis considers historical trends of timber harvesting and sedimentation (i.e., examining "Temporality") in the watershed, yarding activity and roading and the performing of a sediment budget (in which many of the eleven considerations in Table 4-4 appear in a weight-of-evidence approach, including Consistency, Biological gradient, and Plausibility). The state concluded that roading was the greatest source of management-related sediment delivery in the watershed. Loads were calculated accordingly.

4.5.2.2 Little Scioto River, Ohio

A case study presented in U.S. EPA (2000d) illustrates the use of stressor identification criteria for a complicated system in which biological impacts at three sites over a nine-mile stretch were judged to have occurred, possibly caused by several stressors. Although the evaluation compares results from study sites to numeric, rather than narrative, biocriteria, the study is a good example of an assessment using a full set of impairment criteria. Many of the guiding principles presented here were followed in this study (see Table 4-5).

Note that even in this complex system, a single (but different) cause was identified at each site, calling for different load allocations over the entire nine-mile stretch of river. In other cases, there may not be one dominant cause (as noted in U.S. EPA, 2000d), and simultaneous load allocations may be required for different stressors at the same site.

Table 4-5. Implementation of Guiding Principles on the Little Scioto River, Ohio.

Principles	Application to Little Scioto River
Develop list	Six candidate stressors listed: habitat alteration; PAH exposure; metal contamination; ammonia toxicity; low dissolved oxygen; nutrient enrichment.
Conceptual model	Conceptual model flow diagram developed.
Analyze data, refine model	Detailed analyses performed, including nonparametric correlations and assessments of co-location. Criteria for elimination determined.
Eliminate	Some candidate causes were logically eliminated at each of the three sites (three each at sites A and C, only one at site B).
Diagnose	Diagnosis performed, but no strong evidence found to confirm any particular stressor.
Do strength-of-evidence analysis	Completed.
Use full set of criteria	Used 12 criteria.
Weigh the evidence	Scores assigned to each criterion and used to guide the final conclusions. Probable dominant causes were identified: site A, habitat alteration (siltation and channel deepening); site B, PAH-contaminated sediments; site C, increased nutrient enrichment, but evidence less certain. Noted that other causes may also be present.

CHAPTER 5.0

Guiding Principles for Setting TMDL Target Measures and Levels

5.1 Introduction

The previous chapters of this report discussed challenges unique to narrative criteria with respect to defining impairment (Chapter 3.0) and determining the causes of the impairment (Chapter 4.0). Once those steps have been completed, the final challenge unique to developing TMDLs for narrative criteria is setting quantitative TMDL targets.

Establishing TMDL targets based upon narrative criteria is intrinsically difficult because: 1) a quantitative level must be specified based on a qualitative description, and 2) the relevant measures are often undefined in narrative criteria. Furthermore, there is insufficient scientific understanding for many narrative criteria to develop a quantitative cause-effect relationship between stressors and responses. When a complete linkage cannot be defined, the specific measures used for developing the TMDL may differ from the measures used for defining impairment. For example, consider a site where a fishery is impaired by excess sedimentation. Fish population may serve as an appropriate measure for assessing impairment, but is an inappropriate TMDL target measure because of the inability to quantitatively link fish population levels to sediment controls. In this situation, surrogate measures of impairment (e.g., sediment loads) may serve as more appropriate TMDL targets because the relationship between control actions and system response can be estimated directly. In cases where surrogate targets must be used, challenges exist in selecting appropriate target levels to reflect the designated use.

This chapter contains suggested guiding principles to improve the selection of target measures and levels. It is divided into sections:

Challenges and Concerns (Section 5.2)
Selection of TMDL Target Measures (Section 5.3)
Setting TMDL Target Levels (Section 5.4)

5.2 Challenges and Concerns

Three primary challenges were identified in the review of narrative TMDLs related to setting TMDL targets:

- Determining whether to use direct vs. surrogate measures;
- Selecting the specific measure(s) used to reflect the target; and

- Selecting an appropriate target level.

Selection of appropriate measures can differ between the *assessment* phase (determination of impairment) and the *development* phase of the TMDL process, especially when dealing with ecological narrative criteria. Generally, the assessment phase will focus on ecological endpoints and may include quantitative ecological measures of adverse effects. The development phase of the TMDL process may focus on measures of watershed response to alternative control measures. This concept is illustrated in Figure 5-1, which demonstrates the linkage between sources of impairment and environmental response. The parenthetical information in Figure 5-1 contains example measures that represent each link for the case of fish impairment caused by excessive sedimentation.

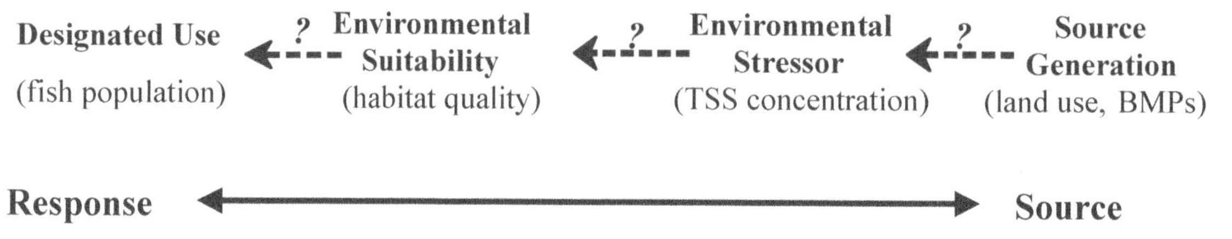

Figure 5-1. Source-Response Relationship, with Example Application of Sediment/Habitat Impairment of Fish Populations.

The far left-hand side of Figure 5-1 represents the environmental response, while the right-hand side represents the sources of impairment. The far right-hand measures relate to the generation of the impairment source (e.g., watershed erosion), as well as to potential means of source control. Moving to the left, the generated source is translated into a stressor level in the environment; for example, eroded solids loads are converted into in-stream TSS concentrations. Next, stressor levels are converted into a measure of environmental suitability, such as habitat quality. Finally, environmental suitability is linked to a direct measure of the designated use.

Each of the arrows in Figure 5-1 represents a predictive relationship. Because the TMDL is designed to result in attainment of the designated use, the selected TMDL target measure should reflect the designated use as closely as possible. In situations where each of the predictive relationships is well defined, the TMDL target can be set at the left-hand side of Figure 5-1, and the target measure may be the same as the measure used to determine impairment. Measures representing the left-hand side of Figure 5-1, which directly represent the designated use, are referred to as direct measures.

The question marks in Figure 5-1 indicate the potential for large degrees of uncertainty in many of the points along this relationship. For example, the relationship between TSS concentrations and resulting habitat quality typically is poorly defined. The TMDL process requires that target measures be quantitatively linked to TMDL controls; setting targets at the response end of the spectrum is difficult when information gaps prevent this quantitative linkage. In such cases, it may be necessary to use measures more closely related to the cause end of the chain (often called "surrogate" measures, as they only indirectly represent the designated use).

The first step in setting TMDL targets consists of selecting the appropriate location on the source-response chain for the target measure (i.e., whether to use direct or surrogate target measures). The second, related step in setting TMDL targets corresponds to selecting the specific measure to be used for the TMDL target. Using Figure 5-1 as an example, the first issue may have been addressed by selecting "Environmental Suitability" as the appropriate location to set targets. A second decision is then required, regarding what specific measure to use for the target (e.g., cobble embeddedness, depth of fines).

The final step is to select the appropriate level (i.e., quantity) for the selected measure. For example, if cobble embeddedness is selected as the target measure, the target level would correspond to the maximum allowable percentage.

The following sections provide guiding principles to address each of the above challenges.

5.3 Selecting Target Measures

The first two challenges in selecting TMDL targets correspond to selection of target measures. The first, broader challenge corresponds to where on the source-response relationship the target measure should be located. The second challenge corresponds to selection of the specific measure to be used. Guiding principles are provided below specific to each of the challenges.

5.3.1 Guiding Principles

Principle #1: Select direct measures of environmental response to serve as TMDL targets when the cause-effect relationship is well understood.

The objective of the TMDL target is to represent attainment of the designated use. The best representation of the designated use will come from a direct measure of the use (i.e., a measure located at the "response" end of the source-response chain). Example situations where direct measures are appropriate include:

- ♦ In the case of a eutrophication impairment in a system with a calibrated eutrophication model, target measures of eutrophication endpoints (e.g., clarity and dissolved oxygen) are more appropriate TMDL targets than measures of nutrient concentration.
- ♦ In the case of a sedimentation impairment related to loss of reservoir capacity, sedimentation rate is an appropriate TMDL target because the relationship between the sedimentation rate and solids load is generally easily defined.
- ♦ In the case of solids impairment when well-established correlations exist between turbidity and TSS concentrations, turbidity is a more appropriate target measure.

This guiding principle includes the qualifier "when the cause-effect relationship is well understood." Principle #2 contains the corollary to this principle, and discusses situations where the cause-effect relationship is less well understood.

Principle #2: Select surrogate measures to serve as TMDL targets when some aspect of the cause-effect relationship is poorly understood.

For many narrative criteria, critical knowledge gaps exist when attempting to develop quantitative cause-effect relationships at some point in the source-response chain. Examples of this include:

- ♦ Insufficient understanding of how in-stream solids concentrations relate to benthic habitat quality.
- ♦ Insufficient understanding of mercury speciation in ambient waterbodies needed to predict ambient levels of bioavailable mercury from total mercury loads.

- Insufficient understanding of the relationship between chlorophyll *a* concentrations and resulting use impairment.
- Insufficient understanding of ecological responses to changes in water temperature.

TMDL target measures must be linked quantitatively to the sources, so that load allocations necessary to meet the targets can be defined. Knowledge gaps can prevent this quantitative linkage between sources and direct measures of impairment, and may require the use of TMDL targets closer to the source end of the relationship. In the case of excess sediment-caused habitat impairment, for example, relationships between land practices and resulting stressor levels (e.g., watershed sediment load) may be well known, but a quantitative relationship between sediment load and habitat quality may not be available. In this case, it is appropriate that the TMDL target measure be representative of the stressor. Guiding principles for selecting specific target measures and levels are provided in subsequent sections.

Principle #3: When a direct measure of environmental response is used as a TMDL target, the measure selected should be the same as (or consistent with) those used to determine impairment.

Selection of the TMDL target measure is relatively straightforward when direct measures of impairment are used. In most cases, the same measure used to determine impairment can be used as the target measure. (Guiding principles for selecting impairment measures were provided in Chapter 3.0, and example measures listed in Table 3-1.) In other cases, however, practical considerations may preclude use of the identical measure used to determine impairment as the target measure (e.g., the measure of impairment cannot be linked quantitatively to the source). In this case, a different measure can be used as the TMDL target as long as it is consistent with the existing impairment measure; that is, the measure is: 1) representative of the same impairment, and 2) equally protective. Consider a lake designated to support primary contact recreation that has been listed as aesthetically impaired by algal blooms, where the impairment measure was defined as peak chlorophyll *a* concentrations. An existing water quality model for the lake can predict the relationship between nutrient loads and summer average chlorophyll concentrations. Peak chlorophyll and summer average chlorophyll are both direct measures of impairment. Summer average chlorophyll can be used as the TMDL target for this case, as long as it can be demonstrated that reductions in summer average chlorophyll will result in equivalent reductions in peak chlorophyll concentrations.

The following example demonstrates a reviewed TMDL that does not comply with this principle. An impoundment on a large river was listed as impaired because of aesthetic impairment caused by algal blooms in the headwaters of the impoundment. In this case, the cause of the impairment was the observation of unacceptable algal blooms, a direct measure of the aesthetic impairment. The source of the impairment was excess nutrient loads to the impoundment. The TMDL target measure selected was dissolved oxygen concentration, because numeric dissolved oxygen standards exist for the site; however, no impairment of dissolved oxygen standards was observed in the impoundment. Dissolved oxygen concentration is an inappropriate TMDL target measure in this case, because it is not fully consistent with the true cause of impairment. Consistency, as defined above, requires that the target measure be: 1) representative of the same impairment, and 2) equally protective. Dissolved oxygen as a target measure was consistent with the first part of the definition because nutrient loads ultimately affect dissolved oxygen concentrations. The target measure fails to satisfy the requirement to be equally protective, however, because nutrient loads necessary to achieve attainment of the dissolved oxygen standard do not assure attainment with aesthetic criteria. A more appropriate target measure for this site would have been some measure of algal biomass, such as chlorophyll *a* concentration.

Principle #4: When multiple direct measures are used to determine impairment, the most protective measure should be used as the TMDL target.

Selection of a direct target measure is simple when only a single measure is used to determine impairment, but becomes more complicated when multiple measures are used. In these cases, each measure should be considered individually, and the most protective measure selected. The analysis of which measure is most protective can in some cases be determined simply, or in other cases may require a full modeling analysis to determine which measure results in the lowest allowable load.

An example of the simple analysis is as follows. A site is listed as impaired because of tissue contamination from two different species of fish, lake trout and perch. Analysis of site-specific bioaccumulation data shows that, for any given water column pollutant concentration, lake trout tissue levels will be more likely to exceed fish tissue target levels than will perch. In this case, lake trout levels can be set as the TMDL target measure without the need for fully deriving separate TMDLs for each target measure.

As an example of more detailed analysis, consider a lake designated to support a warm water fishery that has been listed as ecologically impaired due to an overabundance of blue-green algae. The lake has impairment targets for both peak blue-green algae levels and summer average chlorophyll *a* concentrations. A simple correlation cannot be drawn between these two measures, although models exist that describe the response of each measure to reductions in pollutant loads. In this case, the analysis would need to be conducted independently for each measure to determine which measure requires the greatest nutrient loading reduction to meet the desired level. The measure that requires the most stringent reduction should be selected as the TMDL target.

Principle #5: A surrogate TMDL target measure must be: 1) quantitatively linked to source controls; 2) a significant causative factor determining impairment; and 3) capable of having a target level established corresponding to attainment of the designated use.

Surrogate measures are used for TMDL targets when the linkage between the stressor and the impairment cannot be fully quantified. Common examples occur for biological impairment from excessive sedimentation, where indirect relationships exist between sediment loading and biological integrity. In these instances, the TMDL target measures may include the sediment load itself or one or more of a wide range of sediment reduction strategies aimed at reducing sediment load. Various biological measures and biotic indices may be surrogates for stressors such as sediment and nutrients. Table 5-1 summarizes the range of surrogate measures that have been used in narrative criteria TMDLs.

Table 5-1. Examples of Surrogate Target Measures in Reviewed TMDLs.

Impairment	Surrogate Target Measures
Ecological	watershed areal erosion rate; sediment loads; riparian characteristics; hillslope indicators; fine sediment volume of active bed matrix; percent fines, hydrologic connectivity of roads; diversion potential of roads; stream crossings with high risk of failure; primary pools covering 40% of total habitat length; road density; frequency of required dredging; streambank stability; TSS concentration; percent fines in gravels; percent effective shade
Eutrophication	nutrient concentration; Secchi depth clarity; TSI, nutrient loads; chlorophyll *a*
Toxicological	Whole effluent toxicity

Principle #5 requires that three conditions be satisfied when selecting a surrogate target. First, the quantitative linkage between the target measure and the sources must be defined. This linkage is used to define the capacity of the waterbody to tolerate or assimilate the stressor and support the water's designated uses. Based on this understanding, the allowable loads or load reductions can be allocated among contributing sources.

Second, the surrogate measure must be a significant causative factor determining impairment. Harwell et al. (1999) term this concept "relevance to endpoint." For the TMDL to result in attainment of the designated use, source controls that affect the surrogate stressor must have a corresponding effect on the designated use. Chapter 4.0 provides insight on selecting surrogate measures that satisfy this condition.

The third condition for an acceptable surrogate target measure is that a target level can be established for the measure that corresponds to attainment of the designated use. This condition is typically the most difficult to satisfy, for the following reason. If a direct quantitative relationship existed between the surrogate measure under consideration and a direct measure of the designated use, there would be no need to use a surrogate target measure. Instead, this quantitative relationship would allow the direct measure to serve as the TMDL target. An appropriate surrogate measure is one that can be related to the designated use in a qualitative or semi-quantitative way. Examples include establishing relationships between monitoring measures, such as increasing trends in suspended solids and watershed land development, with adverse effects on biotic health measures. These relationships do not have to provide a complete quantitative description, as long as they identify key effect thresholds (i.e., the level at which changes in the stressor result in nonattainment of the designated use). This final condition for selecting a surrogate measure (i.e., that a target level can be established for the measure) may require feedback from the process of selecting target levels, because definition of target levels is linked to selection of appropriate target measures. It often is necessary to begin the process of target level selection to finalize the selection of the appropriate target measure. (Guiding principles for selecting target levels are provided in Section 5.4.)

5.3.2 Case Example

There are many challenges to finding the relationship between surrogate targets and the environmental endpoints of interest. The Savannah River mercury TMDL (U.S. EPA, 2001b) provides an example of the difficulty of defining target levels. For this TMDL, total mercury was selected as the surrogate measure to protect against bioaccumulation of mercury in fish tissue. Bioaccumulation factors (BAFs) were available to relate methyl mercury concentrations in the water column to fish tissue levels. Site data were used to derive an empirical correlation between total mercury and methyl mercury (Figure 5-2). These data, however, exhibit a weak correlation ($r^2<0.01$). Many factors can contribute to the poor correlation between a stressor and an endpoint. One possibility is incomplete temporal consideration, caused by a time lag between an increase in surrogate level and environmental response. Another possibility is incomplete spatial consideration, with different correlations holding at different locations in the watershed, perhaps driven by locally varying methylation rates. The spatial differences could be caused by the fact that different sources of total mercury throughout the watershed contribute differently to the methyl mercury level.

This example demonstrates the need to examine the quality of the relationship between the target measure and the environmental response, and to determine the cause of any observed variability to the maximum extent possible.

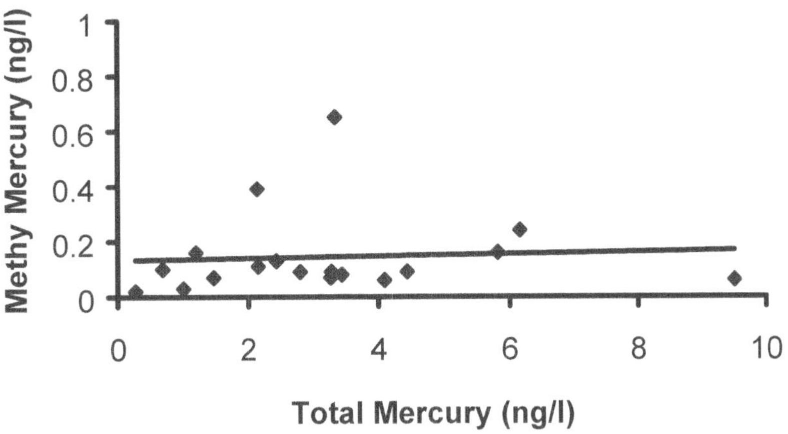

Figure 5-2. Observed Correlation Between Water Column Concentrations of Methyl Mercury and Total Mercury for the Savannah River. (Data are from U.S. EPA, 2001b.)

5.4 Setting Target Levels

Once the TMDL target measure has been identified, the final step in this phase of the TMDL process is to set the specific target level. The guiding principles in this section focus on setting target levels for surrogate measures. When direct measures of impairment are used as TMDL targets, target levels can be set at the same level used for determining impairment. Setting target levels for surrogate measures is more difficult, and is generally based on empirical relationships between observed stressor levels and support of designated uses. These relationships may be obtained from a range of sources, with the primary sources being:

- Site-specific historic monitoring data;
- Data from reference watersheds; and
- Literature relationships from other sites.

5.4.1 Guiding Principles

5.4.1.1 Site Specific Historic Monitoring Data

Principle #1: Site-specific historic monitoring data reflecting stressor levels during periods of designated use attainment provide the most direct means of defining TMDL target levels.

Setting target levels for surrogate measures generally requires knowledge of the relationship between stressor levels and use support that is reflective of the watershed of interest. Historic monitoring data for the watershed may provide guidance in selecting surrogate target levels, if site-specific data for potential surrogate measures are available for a period of time when the designated use was attained.

This approach has been applied successfully in many narrative criteria TMDLs. In one reviewed TMDL, a lake loading target was based on predevelopment water quality measurements. Similar long-term trend relationships may be established for changes in biotic resources with either general watershed development or specific water quality variables. The Noyo River, California, sediment TMDL (U.S. EPA, 1999c) used a target sediment delivery rate from the period 1933-1957, when salmonids were relatively abundant.

The use of historical monitoring data to define stressor-use relationships is not without challenges. Historic data may not be as robust and representative as current data sets. Historic data sets often were not taken to address the issues of a TMDL and may be less representative due to limited sampling frequency and sampling sites, outdated analytical methodologies, and time of sampling. Historic sampling, therefore, may not be directly comparable to current data sets in defining ambient conditions during current time frames. Quality assurance/quality control (QA/QC) information also may be missing for historical data. These issues are all cautions against using historic data that has not been properly documented and studies done to ensure that the historic data methods are comparable to contemporary data methods. Furthermore, in complex biological systems, there is no guarantee that reducing stressor levels to pre-impairment levels will return the system to its prior state of designated use support. Using lake eutrophication as an example, accelerated external phosphorus loading may create a large store of nutrients in lake sediments, such that internal nutrient loads may prevent attainment of designated uses even after reduction of external loads to historical levels.

In the absence of a well-defined cause-effect relationship, however, use of historical stressor levels provides the most straightforward method of target selection. A final consideration is that the historical reference point must reflect conditions that are attainable in the watershed. For example, selection of target levels of maximum allowable impervious surface in a developed watershed to predevelopment levels may not be feasible. In the use of historic data to define TMDL targets, Principles #2 and #3 that follow below for the use of reference watersheds also apply.

5.4.1.2 Reference Watersheds

A second option for determining TMDL target levels is the use of a reference watershed approach. Reference watersheds and reference sites provide a basis for developing cause-effect relationships needed to establish quantitative target levels for TMDL development. The approach uses two watersheds: one that is attaining its uses (the reference watershed) and one that is impaired for which the TMDL must be developed. The focus of this approach is to define the surrogate target level for an impaired segment, based upon the level observed in the reference segment. Reference watersheds are a suitable means to develop target levels when insufficient site-specific monitoring data are available to base target levels on historical data. The reference site-based approach for setting target levels depends on the availability of suitable reference sites, appropriate monitoring data for reference and stressed sites, and the choice of appropriate categories for setting target levels for a given stressor measure. This approach is not without risk; there is no guarantee that the reference target level directly corresponds to attainment of the designated use in the impaired waterbody. It is possible that more lenient or more restrictive target levels will be required to attain the use.

Principle #2: The reference watershed should reflect conditions that are reasonably attainable in the watershed of concern.

This guiding principle corresponds to selection of a reference watershed that possesses conditions appropriate for the TMDL segment. U.S. EPA (1999a) guidance defines reference sites as: "*Waterbodies that are representative of the characteristics of the region and subject to minimal human disturbance.*" This definition is limited in its utility in that it may be interpreted as referring to 'pristine' or a condition before the land was settled. Alternative definitions have been proposed by states, such as:

> 'Reference condition'—a waterbody capable of supporting its present and future beneficial uses when all reasonable land, soil, and water conservation practices have been applied. (Montana DEQ)

This definition is based on the Montana narrative water quality standards which state that "no increases are allowed above naturally occurring conditions of sediment…," where "naturally occurring" refers to "conditions or materials present over which man has no control, or from developed land where 'reasonable' land, soil, and water conservation practices have been applied." This definition recognizes that the reference condition does not have to reflect a pristine state, but rather the best condition that can be achieved given the current use of the watershed.

It often is helpful to consider a number of reference watersheds representing a wide range of ecological conditions, such as pristine, near pristine, best management, and highly developed. This can serve one of two purposes. First, it may allow target levels to be interpolated between existing reference sites in the absence of a reference site that properly reflects the condition of the watershed of concern. Second, this stressor-condition relationship can be used to develop target levels on a regionwide or statewide basis.

Harwell et al. (1999) discuss the relationship between reference conditions, desired conditions, and benchmark conditions (Figure 5-3). Reference conditions are the bounding conditions for the ecosystem environmental conditions, which may have representatives at each end of the spectrum from a high degree of disturbance to a high degree of pristine ecological condition. Desired conditions should be the ultimate goal, that is the target system. Benchmark conditions are intermediate to the desired condition, recognizing that the desired condition may be distant from the current condition. The range of each condition represents the natural variability within a given reference system, as well as other variability associated with natural processes at comparable sites.

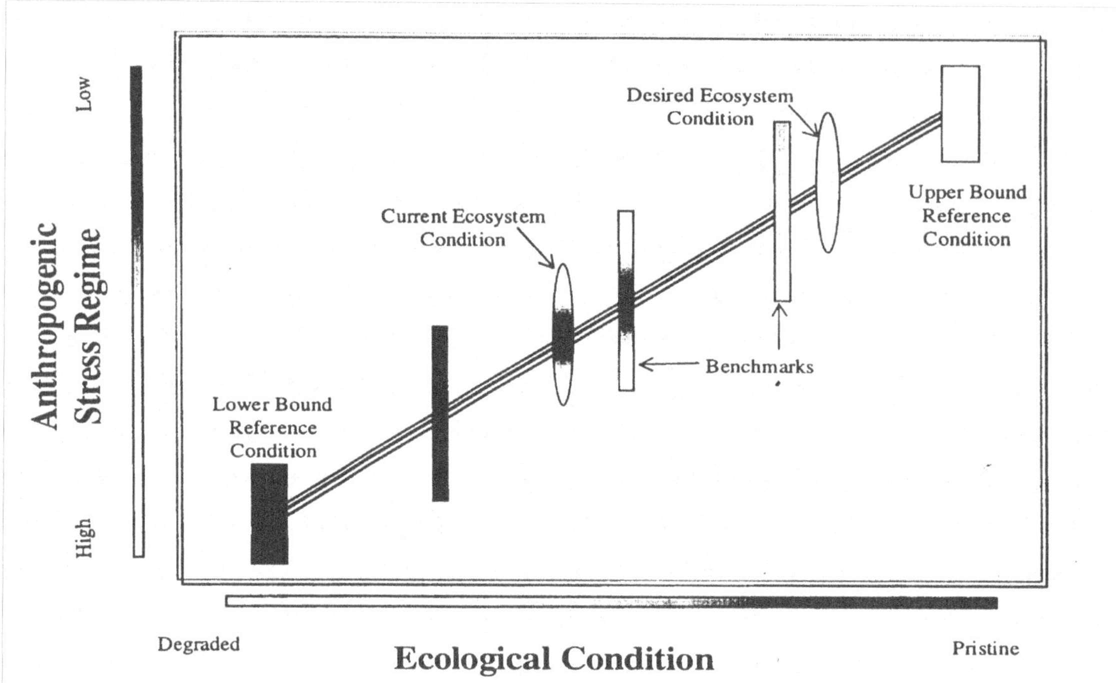

Figure 5-3. Ecosystem Benchmarks for Reference Conditions.
(Reprinted with permission from Harwell et al., 1999. Copyright 1999 American Institute of Biological Sciences.)

Adherence to this principle may require agreement among stakeholders on the selection and appropriateness of the reference condition. This can be conducted following an integrated watershed and ecological risk assessment approach (Serveiss, 2002). The ecological risk assessment process is followed, with the analysis and risk characterization and development of performance criteria for stressor endpoints. These criteria will define what stressor target levels are commensurate with the desired ecological outcome. Analysis of data from a range of reference watersheds may be able to provide a range of benchmarks that are representative of ecosystem health levels. These benchmarks may be used to set interim achievement targets for the TMDL. Additional guidance on the selection of reference conditions is available from U.S. EPA (1990, 1996, 1998c).

5.4.1.3 Literature Target Levels

The final option for selecting corresponding target levels consists of selecting values from the published literature. A large body of scientific literature exists documenting either: 1) stressor levels observed from reference watersheds that have been identified as supporting their designated use for the impairment of concern, or 2) observed relationships between stressor levels and environmental response.

For example, nutrient loads for lakes are commonly calculated on the basis of nutrient/algal biomass models (e.g., Vollenweider, 1976), with selection of the target levels based on the acceptable eutrophication level for the lake. U.S. EPA identifies a number of linkage methods for deriving numeric target levels for sediments, including mathematical linkages, process model linkages, empirical linkages, linkage by inference, and index linkages (U.S. EPA, 1999a).

Examples of other potential literature sources for defining target levels include:

- MacDonald et al., 1991: Provide guidelines to evaluate effects of forestry activities on streams in the Pacific Northwest and Alaska.
- Yoder et al., 2000: Provide correlation between landscape indicators and biological criteria.
- Quinn and Hickey, 1990: Relate pastoral development in forested watershed to reductions in salmonid populations.
- Snelder et al., 2000; Snelder and Biggs, 2002: Provide correlation of nutrients concentrations and biotic responses to land-use activities and hydrologic characteristics.

There are advantages and disadvantages to the use of literature target levels. The advantage is that levels can be found for a wide range of target measures. Their use may be necessary when historical data or data from reference watersheds are not available for the measure of concern. The disadvantage is that literature values pose the greatest potential for misapplication, because the causal mechanisms operating in the published watershed may not reflect conditions in the watershed of interest.

Principle #3: When data from a different watershed (reference or literature relationships) are used to define targets, the causative mechanisms should be similar to those operating in the watershed under consideration.

The use of target levels derived from a watershed other than the one under consideration is limited by the transferability of those numeric levels between ecosystems. For either the reference watershed or the literature approach to work, the watersheds must have similar land cover and land use characteristics. Other features, such as geological (for example drainage area, slope, soil types, etc.), hydrological, and ecological characteristics, should be matched to the maximum extent possible. The difficulties associated with geographic data applicability are common for both ecological assessment (Section 2.5.3) and the subsequent TMDL development. For example, stressors such as nutrient stimulation of planktonic algal abundance in lakes are more widely transferable than nutrient stimulation of attached algae in streams, where hydrologic variability is often a key factor. Similarly, watershed sediment impacts are strongly influenced by a wide range of in-stream and landscape factors, making prediction of impairment more difficult. Thus, literature guidance may provide target levels for some stressors (e.g., nutrients, U.S. EPA, 1999b), but other stressors will require the use of approaches such as classification systems.

In ecological risk assessment, classification schemes serve as a basis for boundary definition in problem formulation and as a tool for extrapolation of exposure-effect relationships in risk characterization (U.S. EPA, 1998d). Existing watershed and landscape classification systems can be classified as either geographically dependent or geographically independent, with further subdivision based on structural or functional ecosystem characteristics (Detenbeck et al., 2000). Utilization of landscape stratification, channel classification and other hierarchical discriminators can provide the technical framework for organizing watersheds. The classification systems can provide direct linkages between factors such as land use and riparian management and stream health measures on a landscape scale (e.g., Detenbeck et al., 2000). Additional guidance on the development and use of classification schemes for surface waters is provided in U.S. EPA (1990, 1996, 1998c).

Several of the TMDLs reviewed used scientific literature as a basis for selecting target values, with varying degrees of appropriateness. The Bear Creek, Idaho, excess sedimentation TMDL used literature relationships between streambank stability and support of designated uses to define a TMDL target of 80% streambank stability (Palisades Sub-basin Assessment and Total Maximum Daily Load Allocations. Donald W. Zaroban and Darcy D. Sharp, State Technical Services Office, Idaho Department of Environmental Quality. January 26, 2001). Tony Tank Lake, Maryland, used a Vollenweider model to define permissible phosphorus loads based on the lake's hydrological and morphological characteristics (Maryland Department of the Environment, 1999).

5.4.2 Case Example

The Newport Bay, California, nutrient TMDL provides a valuable case study demonstrating the use of historic monitoring data to set TMDL target levels (USEPA Region 9, 1999). Newport Bay was listed as impaired because of exceedance of narrative nutrient criteria, based on the presence of large quantities of macroalgae that dominated since the mid-1970s. The relationship between nutrient loading and macroalgal growth was poorly understood at the time of TMDL development, so the use of a direct measure of impairment as a TMDL target was not feasible. However, a large historical database existed describing annual nitrogen load to the Bay, along with observations of algal biomass, which provided qualitative descriptions of the degree of designated use support. The use support data indicated that problems associated with macroalgae began appearing in the mid- to late 1970s. Figure 5-4 shows annual nitrogen loads to the bay, and is divided into two different periods: 1) Period prior to 1977-78 when algae data showed support of the designated use, and 2) Period after 1977-78 when algae data showed non-support of the designated use. The figure demonstrates a large increase in nitrogen loads starting in 1978, coincident in time to the onset of impairment. The TMDL target for Newport Bay nitrogen loading was set at a level of 2×10^6 pounds/year, reflective of loads in the pre-impairment period. This example demonstrates adherence to Principles #2 and #5 for selecting target measures, and Principle #1 for setting target levels. A surrogate measure was selected as a TMDL target because of the lack of a predictive relationship between sources and environmental response (Principle #2 for selecting target measures). The surrogate TMDL target measure selected, annual nitrogen load, was consistent with all three conditions of Principle #5 for selecting target measures: 1) it was quantitatively linked to sources controls; 2) it was a significant causative factor determining impairment; and 3) it was capable of having a target level established corresponding to attainment of the designated use. Finally, selection of a target level was based upon historical data showing nitrogen loads that led to attainment of the designated use, consistent with Principle #1 for setting target levels.

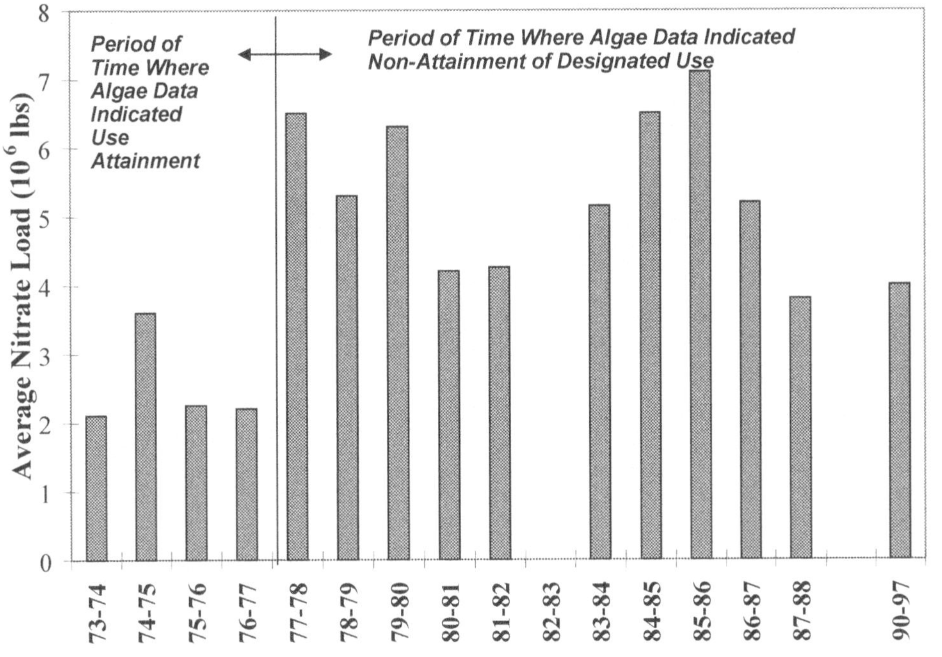

Figure 5-4. Annual Nitrogen Load and Use Attainment Status for Newport Bay, California.
(Data are from California RWQCB, 1997.)

CHAPTER 6.0

CONCLUSION

The objective of this research was to find ways to improve the methods for addressing narrative water quality criteria in the TMDL process. This research compiled and reviewed more than 120 TMDLs covering a wide spectrum of narrative criteria types. The review identified frequent and/or important challenges being in need of guiding principles. Guiding principles were developed for three steps in the TMDL process:

- Defining Impairment
- Determining Causes of Impairment
- Setting TMDL Target Measures and Levels

The guiding principles developed as part of this research are summarized in a set of checklists (Appendix B). Adherence to these guiding principles should significantly improve the evaluation of waterbodies for impairment based on narrative criteria, resulting in more accurate 303(d) lists with a greater focus on those water bodies truly impaired. Also, adherence to these guiding principles will contribute to improvements in the TMDL development process, resulting in a greater degree of success in attaining designated uses when a TMDL is implemented.

APPENDIX A

LIST OF REVIEWED TMDLs

State	Water Body	Narrative Category[1]	Lead Agency	Approval Date	Internet Availability
Alabama	Shades Creek	Ecological	U.S. EPA Region 4	Proposed 10/1/2001	www.epa.gov/region4/water/tmdl/alabama/
Alaska	Duck Creek	Aesthetic	U.S. EPA Region 10	9/26/00	www.state.ak.us/local/akpages/ENV.CONSERV/dawq/tmdl/fin_tmdl.htm
Alaska	Garrison Slough	Toxicological	U.S. EPA Region 10	9/27/96	www.state.ak.us/local/akpages/ENV.CONSERV/dawq/tmdl/fin_tmdl.htm
Alaska	Herring Cove	Ecological	AK Dept. of Env. Conservation	10/1/99	www.state.ak.us/local/akpages/ENV.CONSERV/dawq/tmdl/fin_tmdl.htm
Alaska	King Cove	Ecological	U.S. EPA Region 10	10/9/98	www.state.ak.us/local/akpages/ENV.CONSERV/dawq/tmdl/fin_tmdl.htm
Alaska	Swan Lake	Aesthetic	City and Burough of Sitka, AK	5/25/02	www.state.ak.us/local/akpages/ENV.CONSERV/dawq/tmdl/fin_tmdl.htm
Alaska	Udagak Bay of Beaver Inlet	Ecological	U.S. EPA Region 10	10/9/98	www.state.ak.us/local/akpages/ENV.CONSERV/dawq/tmdl/fin_tmdl.htm
Arizona	Pena Blanca Lake	Toxicological	AZ Dept. of Env. Quality	9/30/99	www.epa.gov/waters/tmdl
Arizona	Stoneman Lake	Aesthetic	AZ Dept. of Env. Quality	3/31/00	www.epa.gov/waters/tmdl
California	Garcia River	Ecological	U.S. EPA Region 9	9/30/97	www.epa.gov/region09/water/tmdl/final.html
California	Navarro River	Ecological & Aesthetic	U.S. EPA Region 9	12/1/00	www.epa.gov/region09/water/tmdl/final.html
California	Noyo River	Ecological	U.S. EPA Region 9	9/30/99	www.epa.gov/region09/water/tmdl/final.html
California	Redwood Creek	Ecological	U.S. EPA Region 9	6/24/98	www.epa.gov/region09/water/tmdl/final.html
California	San Diego Creek and Newport Bay	Ecological	U.S. EPA Region 9/Santa Ana Regional Water Quality Control Board	4/16/99	www.epa.gov/region09/water/tmdl/newport/npnutmdl.pdf
California	South Fork Eel River	Ecological & Aesthetic	U.S. EPA Region 9	12/16/99	www.epa.gov/region09/water/tmdl/final.html
California	South Fork Trinity River	Ecological	U.S. EPA Region 9	11/2/98	www.epa.gov/region09/water/tmdl/final.html
California	Ten Mile River	Ecological	U.S. EPA Region 9	12/27/00	www.epa.gov/region09/water/tmdl/final.html

State	Water Body	Narrative Category[1]	Lead Agency	Approval Date	Internet Availability
California	Van Duzen River and Yager Creek	Ecological	U.S. EPA Region 9	12/16/99	www.epa.gov/region09/water/tmdl/final.html
Colorado	Box Canyon Creek	Ecological	CO Water Quality Control Div.	8/30/00	www.cdphe.state.co.us/wq/Assessment/TMDL/tmdl_status.htm
Colorado	San Miguel River	Ecological	CO Water Quality Control Div.	8/11/00	www.cdphe.state.co.us/wq/Assessment/TMDL/tmdl_status.htm
Delaware	Indian River and Indian River Bay	Aesthetic	DE Dept. of Natural Resources and Env. Control	12/10/98	www.dnrec.state.de.us/water2000/Sections/Watershed/TMDL/tmdlinfo.htm
Delaware	Nanticoke River and Broad Creek	Aesthetic	DE Dept. of Natural Resources and Env. Control	12/10/98	www.dnrec.state.de.us/water2000/Sections/Watershed/TMDL/tmdlinfo.htm
Florida	Lake Thonotosassa	Aesthetic	Florida Dept. of Env. Protection	6/18/98	www.dep.state.fl.us/water/tmdl/status.htm - no TMDL available
Florida	Tampa Bay	Aesthetic	Florida Dept. of Env. Protection	6/18/98	www.dep.state.fl.us/water/tmdl/status.htm - no TMDL available
Georgia	Big Flat Creek	Toxicological	U.S. EPA Region 4	2/8/02	www.epa.gov/region4/water/tmdl/georgia/index.htm
Georgia	Buck Creek	Toxicological	U.S. EPA Region 4	3/7/00	not available
Georgia	Cabin Creek	Toxicological	U.S. EPA Region 4	2/8/02	www.epa.gov/region4/water/tmdl/georgia/index.htm
Georgia	Crawford Creek	Toxicological	U.S. EPA Region 4	3/7/00	not available
Georgia	Jackson Lake	Toxicological	U.S. EPA Region 4	2/28/02	www.epa.gov/region4/water/tmdl/georgia/index.htm
Georgia	Marburg Creek	Toxicological	U.S. EPA Region 4	2/28/02	www.epa.gov/region4/water/tmdl/georgia/index.htm
Georgia	Middle/Lower Chattooga River Watershed	Ecological	U.S. EPA Region 4	4/30/01	www.epa.gov/region4/water/tmdl/georgia/index.htm
Georgia	Mulberry Creek	Ecological	U.S. EPA Region 4	2/1/02	www.epa.gov/region4/water/tmdl/georgia/index.htm
Georgia	Rocky Creek	Toxicological	U.S. EPA Region 4	6/21/01	www.epa.gov/region4/water/tmdl/georgia/index.htm
Georgia	Savannah River	Toxicological	U.S. EPA Region 4	2/28/01	www.epa.gov/region4/water/tmdl/georgia/index.htm
Georgia	Stekoa Creek	Ecological	U.S. EPA Region 4	4/30/01	www.epa.gov/region4/water/tmdl/georgia/index.htm
Georgia	Tired Creek	Ecological	U.S. EPA Region 4	7/2/01	www.epa.gov/region4/water/tmdl/georgia/index.htm
Georgia	Tobesofkee Creek	Toxicological	U.S. EPA Region 4	2/8/02	www.epa.gov/region4/water/tmdl/georgia/index.htm
Georgia	Whites Creek	Toxicological	U.S. EPA Region 4	3/7/00	not available
Idaho	Antelope Creek	Ecological	Idaho Dept. of Env. Quality	2/20/01	www.deq.state.id.us/water/tmdls/tmdls.htm
Idaho	Bear Creek	Ecological	Idaho Dept. of Env. Quality	2/20/01	www.deq.state.id.us/water/tmdls/tmdls.htm

State	Water Body	Narrative Category[1]	Lead Agency	Approval Date	Internet Availability
Idaho	Bruneau River	Aesthetic	Idaho Dept. of Env. Quality	3/13/01	www.deq.state.id.us/water/tmdls/bruneau_river_subbasin/bruneau_river_subbasin_tmdl.htm
Idaho	Jacks Creek	Ecological & Aesthetic	Idaho Dept. of Env. Quality	3/13/01	www.deq.state.id.us/water/tmdls/tmdls.htm
Idaho	Three Creek	Ecological	Idaho Dept. of Env. Quality	3/13/01	www.deq.state.id.us/water/tmdls/tmdls.htm
Idaho/ Washington	Columbia River Basin	Toxicological	U.S. EPA Region 10	2/25/91	yosemite.epa.gov/R10/WATER.NSF/0/062e4bb7e44b8e90882569a700767e8d?OpenDocument
Iowa	Nine Eagles Lake	Aesthetic	IA Dept. of Natural Resources	1/25/01	www.epa.gov/region7/water/apprtmdl.htm
Iowa	Rock Creek	Aesthetic	IA Dept. of Natural Resources	1/25/01	www.epa.gov/region7/water/apprtmdl.htm
Kansas	Clinton Lake	Aesthetic	KS Dept. of Health and Env.	1/26/00	www.kdhe.state.ks.us/tmdl/klrtmdl.htm
Kansas	Kansas River at Lawrence	Toxicological & Aesthetic	KS Dept. of Health and Env.	6/30/99	www.kdhe.state.ks.us/tmdl/klrtmdl.htm
Kansas	Lake Meade State Park	Aesthetic	KS Dept. of Health and Env.	9/7/00	www.kdhe.state.ks.us/tmdl/ciuatmdl.htm
Kansas	Mary's Lake	Ecological	KS Dept. of Health and Env.	1/18/00	www.kdhe.state.ks.us/tmdl/klrtmdl.htm
Kansas	Ogden City Lake	Aesthetic	KS Dept. of Health and Env.	1/18/00	www.kdhe.state.ks.us/tmdl/klrtmdl.htm
Kansas	Soldier Creek Watershed	Ecological & Aesthetic	KS Dept. of Health and Env.	9/7/00	not available
Kansas	Upper Wakarusa River Watershed	Ecological	KS Dept. of Health and Env.	1/26/00	www.kdhe.state.ks.us/tmdl/klrtmdl.htm
Kansas	Warren Park Lake	Aesthetic	KS Dept. of Health and Env.	1/26/00	www.kdhe.state.ks.us/tmdl/klrtmdl.htm
Kansas	Washington WA	Ecological	KS Dept. of Health and Env.	1/26/00	www.kdhe.state.ks.us/tmdl/klrtmdl.htm
Kentucky	Elijahs and Gunpowder Creeks	Ecological	KY Dept. for Env. Protection	2/5/98	water.nr.state.ky.us/dow/tmdl_approved.htm
Kentucky	Taylorsville Lake	Ecological	KY Dept. for Env. Protection	12/8/00	water.nr.state.ky.us/dow/tmdl_approved.htm
Maine	Cobbossee Lake	Aesthetic	ME Dept. of Env. Protection	1/26/00	www.state.me.us/dep/blwq/docmonitoring/tmdl2.htm
Maine	Madawaska Lake	Aesthetic	ME Dept. of Env. Protection	7/24/00	www.state.me.us/dep/blwq/docmonitoring/tmdl2.htm
Maine	Sebasticook Lake	Aesthetic	ME Dept. of Env. Protection	3/8/01	www.state.me.us/dep/blwq/docmonitoring/tmdl2.htm
Maryland	Back River	Toxicological	MD Dept. of Env.	12/17/99	www.mde.state.md.us/Programs/WaterPrograms/TMDL/ApprovedFinalTMDL/index.asp
Maryland	Baltimore Harbor	Toxicological	MD Dept. of Env.	3/20/01	www.mde.state.md.us/Programs/WaterPrograms/TMDL/ApprovedFinalTMDL/index.asp
Maryland	Bohemia River	Aesthetic	MD Dept. of Env.	1/29/01	www.mde.state.md.us/Programs/WaterPrograms/TMDL/ApprovedFinalTMDL/index.asp

State	Water Body	Narrative Category[1]	Lead Agency	Approval Date	Internet Availability
Maryland	Broadford Lake	Aesthetic	MD Dept. of Env.	3/2/00	www.mde.state.md.us/Programs/WaterPrograms/TMDL/ApprovedFinalTMDL/index.asp
Maryland	Chicamacomico River	Aesthetic	MD Dept. of Env.	2/28/01	www.mde.state.md.us/Programs/WaterPrograms/TMDL/ApprovedFinalTMDL/index.asp
Maryland	Corsica River	Ecological	MD Dept. of Env.	3/9/00	www.mde.state.md.us/Programs/WaterPrograms/TMDL/ApprovedFinalTMDL/index.asp
Maryland	Fairlee Creek	Aesthetic	MD Dept. of Env.	3/18/99	www.mde.state.md.us/Programs/WaterPrograms/TMDL/ApprovedFinalTMDL/index.asp
Maryland	Johnson Pond	Aesthetic	MD Dept. of Env.	2/13/01	www.mde.state.md.us/Programs/WaterPrograms/TMDL/ApprovedFinalTMDL/index.asp
Maryland	Lake Habeeb	Aesthetic	MD Dept. of Env.	3/2/00	www.mde.state.md.us/Programs/WaterPrograms/TMDL/ApprovedFinalTMDL/index.asp
Maryland	Port Tobacco River	Aesthetic	MD Dept. of Env.	3/18/99	www.mde.state.md.us/Programs/WaterPrograms/TMDL/ApprovedFinalTMDL/index.asp
Maryland	Tony Tank Lake	Ecological & Aesthetic	MD Dept. of Env.	12/10/99	www.mde.state.md.us/Programs/WaterPrograms/TMDL/ApprovedFinalTMDL/index.asp
Maryland	Transquaking River	Aesthetic	MD Dept. of Env.	3/9/00	www.mde.state.md.us/Programs/WaterPrograms/TMDL/ApprovedFinalTMDL/index.asp
Maryland	Urieville Community Lake	Ecological & Aesthetic	MD Dept. of Env.	8/24/99	www.mde.state.md.us/Programs/WaterPrograms/TMDL/ApprovedFinalTMDL/index.asp
Maryland	Wicomico Creek	Aesthetic	MD Dept. of Env.	1/29/01	www.mde.state.md.us/Programs/WaterPrograms/TMDL/ApprovedFinalTMDL/index.asp
Massachusetts	Bare Hill Pond	Aesthetic	MA Dept. of Env. Protection	11/2/99	www.state.ma.us/dep/brp/wm/tmdls.htm#reports
Michigan	Ore Lake	Aesthetic	MI Dept. of Env. Quality	2/18/00	www.epa.gov/waters/tmdl
Mississippi	Escatawpa River	Toxicological	U.S. EPA Region 4	7/2/01	www.epa.gov/region4/water/tmdl/mississippi/index.htm
Mississippi	Pearl River	Ecological	U.S. EPA Region 4	7/2/01	www.epa.gov/region4/water/tmdl/mississippi/index.htm
Mississippi	Richardson Mill Creek	Ecological	MS Dept. of Env. Quality	6/29/00	www.epa.gov/waters/tmdl
Mississippi	Tidewater Bayou	Toxicological	U.S. EPA Region 4	Proposed 7/2/2001	www.epa.gov/region4/water/tmdl/mississippi/index.htm
Missouri	Clear Creek	Aesthetic	MO Dept. of Natural Resources	12/1/99	www.dnr.state.mo.us/wpscd/wpcp/tmdl/wpc-tmdl-EPA-Appr-date.htm
Missouri	James River	Aesthetic	MO Dept. of Natural Resources	5/7/01	www.dnr.state.mo.us/wpscd/wpcp/tmdl/wpc-tmdl-EPA-Appr-date.htm

State	Water Body	Narrative Category[1]	Lead Agency	Approval Date	Internet Availability
Missouri	North Moreau Creek	Ecological	MO Dept. of Natural Resources	12/1/99	www.dnr.state.mo.us/wpscd/wpcp/tmdl/wpc-tmdl-EPA-Appr-date.htm
Montana	Elk Creek	Ecological	MT Dept. of Env. Quality	12/8/98	www.deq.state.mt.us/ppa/mdm/TMDL/finalReports.asp
Nebraska	Pawnee Reservoir	Ecological	NE Dept. of Env. Quality	3/2/01	www.epa.gov/region7/water/apprtmdl.htm
New Mexico	Rayado Creek	Ecological	NM Env. Dept.	2/16/01	www.nmenv.state.nm.us/swqb/CimarronTMDL.html
North Carolina	Neuse River Estuary	Aesthetic	NC Dept. of Env. & Natural Resources	Proposed 6/21/1999	h2o.enr.state.nc.us/mtu/aboutTMDL.html
Ohio	Middle Cuyahoga River	Ecological	OH Env. Protection Agency	10/11/00	www.epa.state.oh.us/dsw/tmdl/midcuy.html
South Dakota	Lake Hiddenwood	Aesthetic	SD Dept. of Env. & Natural Resources	4/12/99	www.epa.gov/waters/tmdl
South Dakota	Lake Hiddenwood	Aesthetic	SD Dept. of Env. & Natural Resources	4/12/99	www.epa.gov/waters/tmdl
South Dakota	McCook Lake Watershed	Aesthetic	SD Dept. of Env. & Natural Resources	4/12/99	www.epa.gov/waters/tmdl
Texas	Clear Fork Trinity River	Toxicological	TX Natural Resource Conservation Commission	5/24/01	www.tnrcc.state.tx.us/water/quality/tmdl/fwleg.html
Texas	Echo Lake	Toxicological	TX Natural Resource Conservation Commission	5/24/01	www.tnrcc.state.tx.us/water/quality/tmdl/fwleg.html
Texas	Fosdic Lake	Toxicological	TX Natural Resource Conservation Commission	5/24/01	www.tnrcc.state.tx.us/water/quality/tmdl/fwleg.html
Texas	Lake Como	Toxicological	TX Natural Resource Conservation Commission	5/24/01	www.tnrcc.state.tx.us/water/quality/tmdl/fwleg.html
Texas	West Fork Trinity River	Toxicological	TX Natural Resource Conservation Commission	5/24/01	www.tnrcc.state.tx.us/water/quality/tmdl/fwleg.html
Washington	Steilacoom Lake	Aesthetic	U.S. EPA Region 10	2/14/00	yosemite.epa.gov/r10/water.nsf/0/bcfae38a8c193e6088 25673300019295/ $FILE/Steilacoom.pdf
Washington	Lower Yakima River	Ecological	WA Dept. of Ecology	11/25/98	www.ecy.wa.gov/biblio/97321.html
West Virginia	Bear Rocks Lake	Ecological & Aesthetic	U.S. EPA Region 3	9/30/99	www.dep.state.wv.us/alt.cfm?asid=46
West Virginia	Burches Run Lake	Ecological & Aesthetic	U.S. EPA Region 3	10/2/98	not available
West Virginia	Castleman Run Lake	Ecological & Aesthetic	U.S. EPA Region 3	9/30/99	not available

State	Water Body	Narrative Category[1]	Lead Agency	Approval Date	Internet Availability
West Virginia	Hurricane Lake	Ecological & Aesthetic	U.S. EPA Region 3	10/2/98	www.dep.state.wv.us/alt.cfm?asid=46
West Virginia	Mountwood Park Lake	Ecological	U.S. EPA Region 3	10/2/98	www.dep.state.wv.us/alt.cfm?asid=46
West Virginia	Ridenour Lake	Ecological & Aesthetic	U.S. EPA Region 3	9/30/99	www.dep.state.wv.us/alt.cfm?asid=46
West Virginia	Saltlick Pond	Aesthetic	U.S. EPA Region 3	7/5/00	www.dep.state.wv.us/alt.cfm?asid=46
West Virginia	Tomlinson Run Lake	Ecological	U.S. EPA Region 3	10/2/98	www.dep.state.wv.us/alt.cfm?asid=46
West Virginia	Turkey Run Lake	Ecological & Aesthetic	U.S. EPA Region 3	9/30/99	www.dep.state.wv.us/alt.cfm?asid=46

1) Some documents include more than one TMDL, for a total of 120 TMDLs reviewed.

APPENDIX B

CHECKLIST OF GUIDING PRINCIPLES

Checklist for Defining Impairment		
For a Consistent Decision Process		
☐	1	States should develop protocols
☐	2	Consider all relevant available protocols
☐	3	Document all rationale
☐	4	Use an intermediate listing category if impairment is inconclusive
☐	5	Investigate the scientific basis of fish consumption advisories
For Selection of Measures and Levels of Impairment		
☐	1	Link the measures to the designated use
☐	2	Use well-accepted techniques to determine the measure
☐	3	Determine the expected variability of the measurement
☐	4	Measures should be consistent with existing data
☐	5	Measure should be readily understood.
☐	6	Consider multiple measures – weight of evidence approach
☐	7	The threshold level must be relevant to the specific designated use and local condition
For Data Interpretation		
☐	1	Address variability in measurement
☐	2	Consider geographic and temporal relevance
☐	3	Consider all potential causes
For Determining Aesthetic Impairment		
☐	1	Assess potential causes
☐	2	Develop measurement procedures
☐	3	Take into consideration local input

Checklist for Determining Causes of Impairment		
For Listing Candidate Causes		
☐	1	Develop a wide initial list of candidate sources and causes
	2	Develop a conceptual model indicating links between causes and impairment
For Analyzing the Evidence		
☐	1	Analyze available data and literature to refine the conceptual model
☐	2	Assign uncertainties to the model and use the model to guide data collection
For Identifying Potential Causes		
☐	1	Eliminate causes that fail logical tests
☐	2	Subject all remaining causes to diagnostic analysis
☐	3	Perform a strength-of-evidence analysis
☐	4	Apply the Bradford-Hill criteria to the strength-of-evidence analysis
☐	5	Compare the relative strengths of evidence using a weighted matrix for causes

Checklist for Setting TMDL Target Measures and Levels		
For Selecting Target Measures		
☐	1	When cause-effect is well understood select direct measures of environmental response
☐	2	When cause-effect is poorly understood select surrogate measures to serve as TMDL targets
☐	3	A direct TMDL target measure should be the same as used to determine impairment
☐	4	Use the most protective measure as the TMDL target when multiple direct measures are used
☐	5	Surrogate targets must be linked to source controls showing significant causative factors
For Setting Target Levels		
☐	1	Historic monitoring data provide the most direct means of defining TMDL target levels
☐	2	A reference watershed must reflect attainable conditions in the target watershed
☐	3	A reference watershed must have targets with similar causative mechanisms

REFERENCES

Barnes & Thornburg (Law Firm) and Parsons (Consulting Firm). 2002. Preparation of Integrated Water Quality Monitoring and Assessment Reports: Recommendations for Clean Water Act 303(d) and 305(b) Methodologies and Reporting. Volume 1. Federal Water Quality Coalition, The American Farm Bureau Federation, and The Association of Metropolitan Sewerage Agencies.

California RWQCB. 1997. Staff Report on the Nutrient Total Maximum Daily Load for Newport Bay/ San Diego Creed, August 29. 1997. Regional Water Quality Control Board, Santa Ana Region.

Coleman, A.M. 2001. A Dictionary of Psychology. Oxford University Press, New York.

Colorado Department of Health and Environment. 2002. Provisional Implementation Guidance for Determining Sediment Deposition Impacts to Aquatic Life in Streams and Rivers. WQCC Policy 98-1.

Coren, S., C. Porac, and L.M. Ward. 1984. Sensation and Perception. Second Edition. Orlando, Academic Press.

Detenbeck, N.E., S.L. Batterman, V.J. Brady, J.C. Brazner, V.M. Snarski, D.L. Taylor, J.A. Thompson, and J.W. Arthur. 2000. A test of watershed classification systems for ecological risk assessment. Environmental Toxicology and Chemistry 19:1174-1181.

Diamond, J., C. Daley, and T. Moore. 1999. Evaluating Whole Effluent Toxicity as an Indicator of Instream Biological Conditions. Water Environment Research Foundation, Alexandria, VA.

Dinius, S.H. 1981. Public perception in water quality evaluation. Water Resources Bulletin 17: 116-121.

Foran, J.A. and S.A. Ferenc (Eds.). 1999. Multiple stressors in ecological risk and impact assessment. SETAC Pellston workshop on multiple stressors in ecological risk and impact assessments; 13-18 September 1997. Society of Environmental Toxicology and Chemistry (SETAC), Pensacola, Florida.

Fox, G.A. 1991. Practical causal inference for ecoepidemiologists. Journal of Toxicology and Environmental Health 33:359–373.

Germano, J.D. 1999. Ecology, statistics, and the art of misdiagnosis: the need for a paradigm shift. Environmental Reviews 7:167–190.

Goodman, S.N. 1993. P values, hypothesis tests, and likelihood: implications for epidemiology of a neglected historical debate. American Journal of Epidemiology 137:485–496 (discussion on pp. 497–501).

Goodman, S.N. 1999. Toward evidence-based medical statistics. 1: The P value fallacy. Annals of Internal Medicine 130:995–1004.

Griffith, L.M., R.C. Ward, G.B. McBride, and J.C. Loftis. 2001. Data analysis in producing 'comparable' information for water quality management purposes. Technical Report 01-01, National Water Quality Monitoring Council.

Grubbs, G. 2001. Development and Adoption of Nutrient Criteria into Water Quality Standards. Memorandum dated November 14, 2001. U.S. EPA Office of Science and Technology.

Harlow, L.L., S.A. Muliak, and J.H. Steiger. 1997. What if there were no significance tests? Mahwah, NJ, Lawrence Erlbaum Associates. 435 p.

Harwell, M., V. Myers, T. Young, A. Bartuska, N. Gassman, J. Gentile, C. Harwell, S. Appelbaum, J. Barko, B. Causey, C. Johnson, A. McLean, R.T. Smola, P. Templet, and S. Tosini. 1999. A Framework for an Ecosystem Integrity Report Card. BioScience 49:543-556.

Heiskary, S.A. 1989. Lake assessment program: a cooperative lake study program. Lake and Reservoir Management 5:85-94.

Heiskary, S.A. and C.B. Wilson. 1989. The Regional Nature of Lake Water Quality Across Minnesota: An Analysis for Improving Resource Management. Journal of the Minnesota Academy of Science 55 (1):71-77.

Hill, A.B. 1965. The environment and disease: association or causation? Proceedings of the Royal Society of Medicine (Section on Occupational Medicine) 58:295–300.

Karr, J.R. 1981. Assessment of biotic integrity using fish communities. Fisheries 6:21-27.

Karr. J.R and E.W. Chu. 1997. Biological Monitoring and Assessment: Using Multimetric Indexes Effectively. EPA 235-R97-001. University of Washington, Seattle.

Karr, J. R. 2000. Health, integrity, and biological assessment: The importance of whole things. p 209-226 in D. Pimentel, L. Westra and R. F. Noss (eds): Ecological Integrity: Integrating Environment, Conservation, and Health. Island Press, Washington, D.C.

Knopf, R.C. 1983. Recreational needs and behavior in natural settings. In: Altman, I. and Wohlwill, J.F. (eds). Behavior and the Natural Environment. New York, Plenum Press.

Limno-Tech, Inc., Ross & Associates Environmental Consulting, FTN Associates, and Technology Planning and Management Corporation. 2003. Navigating the TMDL Process: Evaluation and Design, A WERF Project Update. Final Report prepared for Water Environment Research Foundation, Project 00-WSM-1, expected publication 2003.

MacDonald, L.H., A.W. Smart, and R.C. Wissmar. 1991. Monitoring guidelines to evaluate effects of forestry activities on streams in the Pacific Northwest and Alaska. U.S. Environmental Protection Agency, EPA/910/9-9-001, Washington, D.C.

MacKenzie, W.R., N.J. Hoxie, M.E. Proctor, M.S. Gradus, K.A. Blair, D.E. Peterson, J.J. Kazmierczak, K.R. Fox, D.G. Addias, J.B. Rose, and J.P. Davis. 1994. A massive outbreak in Milwaukee of *Cryptosporidium* infection transmitted through the public water supply. New England Journal of Medicine 331(3):161–167.

Maryland Department of the Environment. 1999. Total Maximum Daily Loads of Phosphorus and Sediments to Tony Tonk Lake, Wicomico County, MD.

McBride, G.B. 1999. Equivalence tests can enhance environmental science and management. Australian & New Zealand Journal of Statistics 41(1):19–29.

McBride, G.B. 2002. Statistical methods helping and hindering environmental science and management. Journal of Agricultural, Biological and Environmental Statistics 7(3):300-305.

NALMS, 1992. Developing Eutrophication Standards for Lakes and Reservoirs. A Report Prepared for the Lake Standards Subcommittee. North American Lake Management Society, Alachua, FL.

Nelder, J.A. 1999. Statistics for the millennium: from statistics to statistical science. The Statistician 48(2):257–269.

North Carolina Department of Environment and Natural Resources. 2000. Appendix II: Determining Use Support for Eutrophication Related Water Quality Variables. North Carolina's 2000 § 303(d) List. Division of Water Quality. April 3, 2000 (Final Draft Submitted to EPA).

NRC. 2001. Assessing the TMDL approach to water quality management. Water Science and Technology Board, Division of Earth and Life Studies, National Research Council. Washington D.C. 82 p.

Platt, J.R. 1964. Strong inference. Science 146:347-353.

Popper, K.R. 1968. The Logic of Scientific Discovery. New York, Harper & Row.

Pott, D.B., D.L. Mulvey, and B.J. Yurdin. 2002. Developing TMDLs in the Absence of Numeric Standards. Proceeding of the Watershed 2002 Conference, Water Environment Federation, Alexandria, VA.

Quinn, J.M. and C.W. Hickey. 1990. Magnitude of effects of substrate particle size, recent flooding, and catchment development on benthic invertebrates in 88 New Zealand rivers. New Zealand Journal of Marine and Freshwater Research 24:411-427.

Quinn, J.M. and W.J. Walley. 2002. Modelling forestry effects on stream ecosystems using a Bayesian Belief Network (Abstract). Bulletin of the North American Benthological Society 19(1):173.

Rand, G.M. (Ed.) 1995. Fundamentals of Aquatic Toxicology: Effects, Environmental Fate, and Risk Assessment, Second Edition. Taylor and Francis, Washington, D.C. 1125pp.

Royall, R. 1997. Statistical Evidence: A Likelihood Paradigm. London: Chapman and Hall.

Serveiss, V.B. 2002. Applying Ecological Risk Principles to Watershed Assessment and Management. Environmental Management 29:145-154.

Smeltzer, E. and S.A. Heiskary. 1990. Analysis and Applications of Lake User Survey Data. Lake and Reservoir Management, 6:109-118.

Smith, D.G., G.F. Croker, and K. McFarlane. 1995. Human perception of water appearance. 2. Colour judgment, and the influence of perceptual set on perceived water suitability for use. New Zealand Journal of Marine and Freshwater Research 29:45-50.

Snelder, T., B.J.F. Biggs, and M. Weatherhead. 2000. Inclusion of Ecosystem Concepts in Integrated Management of River Resources. International Symposium on Integrated Water Resources Management, Davis, California, pp. 379-385.

Snelder, T. and B.J.F. Biggs. 2002. Multi-Scale River Environment Classification for Water Resources Management. Journal of the American Water Resources Association. In press.

Susser, M. 1986. Rules of inference in epidemiology. Regulatory Toxicology and Pharmacology 6:116–128.

Thornton, K.W. and P.L. Freedman. 2003. Navigating the TMDL Process: Listing and Delisting. Water Environment Research Foundation Report, 00-WSM-2, Alexandria, VA.

USDA. 2000. Evaluation of Sediment Transport Data for Clean Sediment TMDLs. National Sedimentation Laboratory. NSL Report No. 17.

U.S. Department of Health, Education and Welfare. 1964. Smoking and Health. Report of the Advisory Committee to the Surgeon General, Washington D.C. Public Health Service Publication 1103.

U.S. EPA. 1974. The Relationship of Phosphorus and Nitrogen to the Trophic State of Northeast and North Central Lakes and Reservoirs. Working Paper No. 23, National Eutrophication Survey, Corvallis, OR.

U.S. EPA. 1976. Quality Criteria for Water [The Red Book]. EPA Number 440976023. U.S. Environmental Protection Agency, Washington, D.C.

U.S. EPA. 1986. Quality Criteria for Water [The Gold Book]. EPA Number 440586001. U.S. Government Printing Office, Washington, D.C.

U.S. EPA. 1989. Rapid Bioassessment Protocols for Use in Streams and Rivers. EPA/444/4-89-001. U. S. Environmental Protection Agency, Washington, D.C.

U.S. EPA. 1990. Biological Criteria: National Program Guidance for Surface Waters. EPA-440/5-90-004. U.S. Environmental Protection Agency, Washington, D.C.

U.S. EPA. 1991. Technical Support Document for Water Quality-based Toxics Control. Office of Water. EPA/505/2-90-001, PB91-127415. U.S. Environmental Protection Agency, Washington, D.C.

U.S. EPA. 1994. Water Quality Standards Handbook, Second Edition, EPA 823-B-94-005a. U.S. Environmental Protection Agency, Washington, D.C.

U.S. EPA. 1996. Biological Criteria: Technical Guidance for Streams and Small Rivers, Revised Edition. EPA/822/B-96/001. U.S. Environmental Protection Agency, Washington, D.C.

U.S. EPA. 1998a. 1998 Section 303(d) List Fact Sheet: Michigan. http://www.epa.gov/owow/tmdl/states/mifact.html

U.S. EPA. 1998b. Report of the Federal Advisory Committee on the TMDL Program. EPA 100-R-98-006. U. S. Environmental Protection Agency, Washington, D.C.

U.S. EPA. 1998c. Guidelines for Ecological Risk Assessment. EPA/630/R-95/002F. U.S. Environmental Protection Agency, Risk Assessment Forum. Washington, D.C. 175 pp

U.S. EPA. 1998d. Lake and Reservoir Bioassessment and Biocriteria: Technical Guidance Document EPA 841-B-98-007. U.S. Environmental Protection Agency, Office of Water, Washington, D.C.

U.S. EPA. 1999a. Protocol for Developing Sediment TMDLs, First Edition, EPA 841-B-99-004. U. S. Environmental Protection Agency, Washington, D.C.

U.S. EPA. 1999b. Protocol for developing nutrient TMDLs. EPA 841-B-99-007. United States Environmental Protection Agency, Office of Water, Washington, D.C.

U.S.EPA. 1999c. Noyo River Total Maximum Daily Load for Sediment. U.S. Environmental Protection Agency, Region 9, San Francisco, CA. December 16, 1999.

U.S. EPA. 2000a. Revisions to the Water Quality Planning and Management Regulation and Revisions to the National Pollutant Discharge Elimination System Program in Support of Revisions to the Water Quality Planning and Management Regulation, Final Rules.

U.S. EPA. 2000b. Use of Fish and Shellfish Advisories and Classifications in 303(d) and 305(b) Listing Decisions. Memorandum from Geoffrey Grubbs, Office of Science and Technology, and Robert Wayland, Office of Wetland, Oceans, and Watersheds. Oct. 24, 2000).

U.S. EPA. 2000c. Evaluation Guidelines for Ecological Indicators. EPA/620/R-99/005. U.S. Environmental Protection Agency, Washington, D.C.

U.S. EPA. 2000d. Stressor Identification Guidance Document. Office of Research and Development. Washington D.C. EPA/822/B-00/025.

U.S. EPA. 2001a. Draft Consolidated Assessment and Listing Methodology: Toward a Compendium of Best Practices. Office of Wetlands, Oceans and Watersheds. April 20, 2001.

U.S. EPA. 2001b. Total Maximum Daily Load (TMDL) for Total Mercury in Fish Tissue Residue in the Middle/Lower Savannah River Watershed, GA. U.S. Environmental Protection Agency, Region 4, Atlanta, GA. February 28, 2001.

U.S. EPA. 2002a. Nutrient Criteria Development, Notice of Ecoregional Nutrient Criteria. Federal Register: February 28, 2002 (Volume 67, Number 40).

U.S. EPA. 2002b. Setting Ecologically Based Water Quality Goals. http://www.epa.gov/waterscience/biocriteria/casestudies/ecobased.html

U.S. EPA. 2002c. Clinch and Powell Valley watershed ecological risk assessment. EPA/600/R-01/050. U.S. Environmental Protection Agency, Office of Research and Development, National Center for Environmental Assessment. Washington, D.C.

U.S. EPA Region 9. 1999. Total Maximum Daily Load for Nutrients: San Diego Creek and Newport Bay, California.

Vollenweider, R.A. 1976. Advances in Defining Critical Loading Levels for Phosphorus in Lake Eutrophication. Mem. Ist. Ital. Idrobiol. 33:53-83.

Washington Department of Ecology. 1997. Assessment of Water Quality for the Section 303(d) List. WQP Policy 1-11.

Water Quality Control Division. 2002. Year 2002 303(d) Listing Methodology. State of Colorado. March 11, 2002.

Yoder, C.O. and E.T. Rankin. 1995. Biological criteria program development and implementation in Ohio, pp. 109-144 (Chapter 9). In W.S. Davis and T. Simon (eds.). Biological Assessment and Criteria: Tools for Water Resource Planning and Decision Making. Lewis Publishers, Boca Raton, FL.

Yoder, C.O., R.J. Miltner, and D.A. White. 2000. Using Biological Criteria to Assess and Classify Urban Streams and Develop Improved Landscape Indicators, pp 32-44. In S. Minamyer et al. (eds.) National Conference on Tools for Urban Water Resource Management and Protection. U.S. EPA, Office of Research and Development, Cincinnati, OH EPA/625/R-00/001.

WASTEWATER UTILITY

Alabama
Montgomery Water Works & Sanitary Sewer Board

Alaska
Anchorage Water & Wastewater Utility

Arizona
Gila Resources

Glendale, City of, Utilities Department

Mesa, City of

Phoenix Water Services Department

Pima County Wastewater Management

Arkansas
Little Rock Wastewater Utility

California
Calaveras County Water District

Central Contra Costa Sanitary District

Contra Costa Water District

Crestline Sanitation District

Delta Diablo Sanitation District

Dublin San Ramon Services District

East Bay Municipal Utility District

El Dorado Irrigation District

Escondido, City of

Fairfield-Suisun Sewer District

Irvine Ranch Water District

Las Virgenes Municipal Water District

Lodi, City of

Los Angeles, City of

Los Angeles County, Sanitation Districts of

Napa Sanitation District

Orange County Sanitation District

Palo Alto, City of

Riverside, City of

Sacramento Regional County Sanitation District

San Diego Metropolitan Wastewater Department, City of

San Francisco, City & County of

San Jose, City of

Santa Barbara, City of

Santa Rosa, City of

South Bayside System Authority

South Orange County Wastewater Authority

Union Sanitary District

West Valley Sanitation District

Colorado
Boulder, City of

Colorado Springs, City of

Littleton/Englewood Water Pollution Control Plant

Metro Wastewater Reclamation District, Denver

Connecticut
New Haven, City of, WPCA

District of Columbia
District of Columbia Water & Sewer Authority

Florida
Broward, County of

Fort Lauderdale, City of

Gainesville Regional Utilities

JEA

Kissimmee, City of, Department of Water Resources

Miami-Dade Water & Sewer Authority

Orange County Utilities Department

Orlando, City of

Reedy Creek Improvement District

Seminole County Environmental Services

St. Petersburg, City of

Stuart Public Utilities

Tallahassee, City of

Tampa, City of

West Palm Beach, City of

Georgia
Atlanta Department of Watershed Management

Augusta, City of

Clayton County Water Authority

Cobb County Water System

Columbus Water Works

Fulton County

Gwinnett County Department of Public Utilities

Macon Water Authority

Savannah, City of

Hawaii
Honolulu, City and County of

Idaho
Pocatello, City of

Illinois
American Bottoms Wastewater Treatment Plant

Dupage, County of

Greater Peoria Sanitary District

Kankakee River Metropolitan Agency

Metropolitan Water Reclamation District of Greater Chicago

North Shore Sanitary District

Wheaton Sanitary District

Indiana
Fort Wayne, City of

Iowa
Ames, City of

Des Moines Metro Wastewater Reclamation Authority

Kansas
Johnson County Unified Wastewater Districts

Unified Government of Wyandotte County/Kansas City, City of

Kentucky
Louisville & Jefferson County Metropolitan Sewer District

Louisiana
Sewerage & Water Board of New Orleans

Maine
Bangor, City of

Portland Water District

Maryland
Anne Arundel County Bureau of Utility Operations

Howard County Department of Public Works

Washington Suburban Sanitary Commission

Massachusetts
Boston Water & Sewer Commission

Upper Blackstone Water Pollution Abatement District

Michigan
Ann Arbor, City of

Detroit, City of

Grand Rapids, City of

Holland Board of Public Works

Lansing, City of

Owosso Mid-Shiawassee County WWTP

Saginaw, City of

Wayne County Department of Environment

Wyoming, City of

Minnesota
Rochester, City of

Missouri
Independence, City of

Kansas City Missouri Water Services Department

Little Blue Valley Sewer District

Metropolitan St. Louis Sewer District

Nebraska
Lincoln Wastewater System

Nevada
Henderson, City of

New Jersey
Passaic Valley Sewerage Commissioners

New York
New York City Department of Environmental Protection

North Carolina
Cary, Town of

Charlotte/Mecklenburg Utilities

Durham, City of

Metropolitan Sewerage District of Buncombe County

Orange Water & Sewer Authority

Ohio
Akron, City of

Butler County Department of Environmental Services

Columbus, City of

Metropolitan Sewer District of Greater Cincinnati

Northeast Ohio Regional Sewer District

Summit, County of

Oklahoma
Tulsa, City of

Oregon
Clean Water Services

Eugene/Springfield Water Pollution Control

Water Environment Services

Pennsylvania
Philadelphia, City of

University Area Joint Authority, State College

South Carolina
Charleston Commissioners of Public Works

Mount Pleasant Waterworks & Sewer Commission

Spartanburg Sanitary Sewer District

Tennessee
Cleveland, City of
Knoxville Utilities Board
Murfreesboro Water & Sewer Department
Nashville Metro Water Services

Texas
Amarillo, City of
Austin, City of
Dallas Water Utilities
Denton, City of
El Paso Water Utilities
Fort Worth, City of
Gulf Coast Waste Disposal Authority
Houston, City of
San Antonio Water System
Trinity River Authority

Utah
Salt Lake City Corporation

Virginia
Alexandria Sanitation Authority
Arlington, County of
Fairfax County Virginia
Hampton Roads Sanitation District
Henrico, County of
Hopewell Regional Wastewater Treatment Facility
Loudoun County Sanitation Authority
Lynchburg Regional WWTP
Prince William County Service Authority
Richmond, City of

Washington
Edmonds, City of
Everett, City of
King County Department of Natural Resources
Seattle Public Utilities
Yakima, City of

Wisconsin
Green Bay Metro Sewerage District
Madison Metropolitan Sewerage District
Milwaukee Metropolitan Sewerage District
Racine, City of
Sheboygan Regional Wastewater Treatment
Wausau Water Works

Australia
South Australian Water Corp.
Sydney Water Corp.
Water Corp. of Western Australia

Canada
Greater Vancouver Regional District
Toronto, City of, Ontario

Mexico
Servicios de Agua y Drenaje de Monterrey, I.P.D.

United Kingdom
Yorkshire Water Services Limited

STORMWATER UTLITY

California
Los Angeles, City of, Department of Public Works
Monterey, City of
Santa Rosa, City of

Colorado
Boulder, City of

Georgia
Griffin, City of

Kansas
Overland Park, City of

Kentucky
Louisville & Jefferson County Metropolitan Sewer District

Maine
Portland Water District

North Carolina
Cary, Town of
Charlotte, City of, Stormwater Services

Pennsylvania
Philadelphia, City of

STATE

Fresno Metropolitan Flood Control District, Calif.
Urban Drainage & Flood Control District, Colo.

CORPORATE

ADS Environmental Services
The ADVENT Group Inc.
Alan Plummer & Associates
Alden Research Laboratory
Alpine Technology Inc.
Aquateam–Norwegian Water Technology Centre A/S
BaySaver Inc.
BioVir Inc.
Black & Veatch
Boyle Engineering Corporation
BPR CSO
Brown & Caldwell
Burns & McDonnell
CABE Associates Inc.
The Cadmus Group
Camp Dresser & McKee Inc.
Carollo Engineers Inc.
Carpenter Environmental Associates Inc.
CDS Technologies Inc.
Chemtrac Systems Inc.
CH2M HILL
Clancy Environmental Consultants Inc.
Damon S. Williams Associates, LLC
David L. Sheridan, P.C.
Earth Tech Inc.
Ecolab Water Care Services
EMA Inc.
The Eshelman Company Inc.
Finkbeiner, Pettis, & Strout Inc.
Frontier Geosciences Inc.
ftn Associates Inc.
Gannett Fleming Inc.
GE Betz
Golder Associates Inc.
Greeley and Hansen LLC
The HACH Company
Hazen & Sawyer, P.C.
HDR Engineering Inc.
HNTB Corporation
HydroQual Inc.
Insituform Technologies Inc.
Institute for Environmental Technology & Industry, Korea
Jacobson Helgoth Consultants Inc.
Jason Consultants Inc.
Jordan, Jones, & Goulding Inc.
KCI Technologies Inc.
Kelly & Weaver, P.C.
Kennedy/Jenks Consultants
Komline Sanderson Engineering Corporation
Lawler, Matusky & Skelly Engineers, LLP
Limno-Tech Inc.
Lombardo Associates Inc.
Malcolm Pirnie Inc.
McKim & Creed
MEC Analytical Systems Inc.
MWH
New England Organics
Odor & Corrosion Technology Consultants Inc. (OCTC)
ONDEO Degremont Inc.
Oswald Green, LLC
PA Government Services Inc.
Parametrix Inc.
Parsons
Post, Buckley, Schuh & Jernigan
R&D Engineering/Conestoga Rover & Associates
The RETEC Group
R.M. Towill Corporation
Ross & Associates Ltd.
Royce Technologies
SAIC Maritime Technical Group
Sear-Brown
Sierra Environmental Services
SYNAGRO
Tetra Tech Inc.
Trojan Technologies Inc.
URS Corporation
USFilter NATC
Vortechnics Inc.
Wade-Trim Inc.
Weston Solutions Inc.
Woodard & Curran
WRc/D&B, LLC
WWETCO, LLC

INDUSTRY

American Electric Power
ChevronTexaco Energy Research & Technology Company
The Coca-Cola Company
Dow Chemical Company
DuPont Company
Eastman Kodak Company
Merck & Company Inc.
ONDEO Services
Procter & Gamble Company
PSEG Services Corp.
Severn Trent Services Inc.
Shell Global Solutions
Thames Water Plc
United Water Services LLC

Note: List as of 5/12/03

Board of Directors

Research Council

Stormwater Technical Advisory Committee

IWA Publishing's authorised EU representative for General Product Safety Regulations is Diane D'Arras, 15 rue Duret, 75116 Paris, France, e-mail: safety@iwap.co.uk.

Printed and bound by CPI Group (UK) Ltd, Croydon, CR0 4YY
12/05/2026
02108879-0009